DRYSTONE RETAINING WALLS

Design, Construction and Assessment

Applied Geotechnics Series

Series Editor: William Powrie
University of Southampton, United Kingdom

PUBLISHED TITLES

David Muir Wood, *Geotechnical Modelling*
Hardback ISBN 978-0-415-34304-6 • Paperback ISBN 978-0-419-23730-3

Alun Thomas, *Sprayed Concrete Lined Tunnels*
Hardback • ISBN 978-0-415-36864-3

David Chapman *et al.,* *Introduction to Tunnel Construction*
Hardback ISBN 978-0-415-46841-1 • Paperback ISBN 978-0-415-46842-8

Catherine O'Sullivan, *Particulate Discrete Element Modelling*
Hardback • ISBN 978-0-415-49036-8

Steve Hencher, *Practical Engineering Geology*
Hardback ISBN 978-0-415-46908-1 • Paperback ISBN 978-0-415-46909-8

Martin Preene *et al.,* *Groundwater Lowering in Construction*
Hardback • ISBN 978-0-415-66837-8

Steve Hencher, *Practical Rock Mechanics*
Paperback • ISBN 978-1-4822-1726-1

Mike Jefferies *et al.,* *Soil Liquefaction, 2nd ed*
Hardback • ISBN 978-1-4822-1368-3

Paul F. McCombie *et al.,* *Drystone Retaining Walls: Design, Construction and Assessment*
Hardback • ISBN 978-1-4822-5088-6

FORTHCOMING

Zixin Zhang *et al.,* *Fundamentals of Shield Tunnelling*
Hardback • ISBN 978-0-415-53597-7

Christoph Gaudin *et al.,* *Centrifuge Modelling in Geotechnics*
Hardback • ISBN 978-0-415-52224-3

Kevin Stone *et al.,* *Weak Rock Engineering Geology and Geotechnics*
Hardback • ISBN 978-0-415-56071-9

DRYSTONE RETAINING WALLS

Design, Construction and Assessment

Paul F. McCombie

University of Bath, United Kingdom

Jean-Claude Morel

LGCB ENTPE, Vaulx en velen cedex, France

Denis Garnier

*Laboratoire Navier, Ecole Nationale des Ponts et Chaussées,
Marne la Vallée Cedex, France*

CRC Press
Taylor & Francis Group
Boca Raton London New York

CRC Press is an imprint of the
Taylor & Francis Group, an **informa** business

A SPON PRESS BOOK

Authors' warnings: Neither the authors nor the publishers of this book can accept any liability in connection with its use. It is the responsibility of the user to ensure that calculations, checks and safeguards are followed in accordance with good engineering practice, with the requirements of the territories from which and in which they operate, and with their own good judgement. The practice of civil engineering and geotechnical engineering especially, requires that a careful assessment is made of any site, and its relationship with its surroundings, as well as of the work being assessed or proposed. Carefully considered advice and recommendations are given in this book, which may inform decisions, but all decisions are the responsibility of the user, not of the authors nor of the publishers.

CRC Press
Taylor & Francis Group
6000 Broken Sound Parkway NW, Suite 300
Boca Raton, FL 33487-2742

First issued in paperback 2019

© 2016 by Taylor & Francis Group, LLC
CRC Press is an imprint of Taylor & Francis Group, an Informa business

No claim to original U.S. Government works

ISBN-13: 978-1-4822-5088-6 (hbk)
ISBN-13: 978-0-367-87040-9 (pbk)

Visit the Taylor & Francis Web site at
http://www.taylorandfrancis.com

and the CRC Press Web site at
http://www.crcpress.com

We dedicate this book to our wives:
Heather McCombie, Rabia Morel and Séverine Garnier.

Contents

Preface

When I was a teenager my family went on annual holidays to Perthshire, in Scotland, where one of my favourite activities was making jetties out of stone on the shores of lochs; less than a metre high, projecting a few metres into the water, these were not great engineering structures, but they gave me a sense of how stone can be fitted together to give a stable and durable structure that might survive the winter storms. In later life, I became very involved in reinforced soil retaining walls, which are very durable and sustainable structures. Later at Bath, Pete Walker asked me to take a look at some modelling work he had done on drystone structures, and thus I was drawn into what has proved to be a very interesting field of study. Pete had already been working with Jean-Claude Morel, who was doing pioneering full-scale testing on structures loaded by water-filled bags, and when we both embarked on full-scale testing of walls loaded by gravel, we kept a keen interest in each other's work, exchanging visits. By this time Denis Garnier had become involved, and we three met for the first time at Jean-Claude's first test in Le Pont de Montvert, in the Cévennes. This was also where I met the brilliant engineers and wallers of Artisans Bâtisseurs en Pierres Sèches (ABPS) and their Directrice, Cathie O'Neill. With my colleagues Kevin Briggs and Laura Warren, I was privileged to spend two days with Thomas Brasseur and Bruno Durand in September 2013, learning more about their art, being shown some of their excellent work, and that of Roland Mousquès, a founder member of the Association, and an inspirational and artistic builder. The work of ABPS, more than anything else, has inspired this book; Chapter 5 draws heavily from their newly prepared 'règles professionnelles'. Thus, a key aim of this work has been to communicate the highly developed French practice of drystone retaining wall construction to a broader audience.

Paul F. McCombie

Acknowledgements

We thank our wives, to whom this book is dedicated, for their patience as we have worked slowly through the task of preparing this book, alongside the many other tasks that crowd the lives of academics. We also thank Tony Moore, of Taylor & Francis, without whose patience and perseverance this book would never have appeared.

The work presented here is in many cases the result of collaborations with colleagues, whose support and encouragement is acknowledged with gratitude. For Paul, some of these are Pete Walker, Andrew Heath, William Powrie, John Harkness, Richard Harkness, Chris Mundell, Sophie Hayward, Richard Tufnell, Laura Warren, Emily Dixon, Shane Donohue, Kevin Briggs, Charlotte Blood, Elaine Roberts, Bruno Durand, Thomas Brasseur, Anne-Sophie Colas, Hanh Le, Eric Vincens, Patrick de Buhan and James Oetomo. Others, too many to list, have supported the work in one way or another, but these are the colleagues whose work or discussion has been most significant – in addition, of course, to his co-authors! The full-scale testing carried out at the University of Bath work was supported by the UK Engineering and Physical Sciences Research Council Grant number EP/D037565/1.

Jean-Claude and Denis thank Paul for leadership in writing this book. They also thank Anne-Sophie Colas, Hanh Le, Boris Villemus, Bruno Durand, Thomas Brasseur, Eric Vincens, Patrick de Buhan, James Oetomo, Guillaume Habert, Ali Mesbah, Claude Boutin, Odile Roque, Jean-François Halouze, Sébastien Courrier, Erwan Hamard, Stéphane Cointet, Joachim Blanc-Gonnet, Paul Arnaud, Philippe Alexandre, Yanick Tardivel, Bertrand Thidet, Roland Somda, Gilbert Haiun, the Artisans Bâtisseurs en Pierres Sèches (Drystone builders' association), in particular Marc Dombre, Christian Emery, Bruno Durand, Thomas Brasseur and Cathie O'Neill. Jean-Claude Morel was encouraged by Patrick Cohen to work on drystone in 1998.

Research in France was also done in the context of two national projects: PEDRA and RESTOR:

1. Réseau Génie Civil et Urbain (RGCU) PEDRA project no. 10 MGC S 017, studies on drystone or weak masonry of the Civil and Urban Engineering Network, coordinated by Eric Vincens of the Ecole Centrale de Lyon.
2. RESTOR project, restoration of dry stone retaining structures, of the Programme National de Recherche sur la Connaissance et la Conservation du patrimoine culturel matériel (PNRCC) program of the Ministry of Culture and Communication, coordinated by Eric Vincens of the Ecole Centrale de Lyon.

Finally, Jean-Claude Morel was supported by the Région Rhône-Alpes in furthering his studies in England, with a 5-month placement at the University of Bath in 2013.

Authors

Paul F. McCombie graduated with a BA in Engineering from Cambridge University in 1981. For the next five years he worked for a consulting engineer, mostly on geotechnical aspects of major road schemes, and studied for an MSc in Soil Mechanics at Imperial College, London. Three years with Netlon Limited (now Tensar International) working on the development of reinforced soil were followed by a move to the University of Bath at the start of 1990, taking over all of the geotechnical engineering lecturing. He has been Director of Studies for civil engineering and Head of Civil Engineering. He has served as Deputy Head of Department since 2009, when he was awarded his PhD on the basis of published research in geotechnical engineering. This work included a new method for slope stability analysis based on displacement, which made use of his earlier work using genetic algorithms to find critical mechanisms. His research on drystone retaining walls within the BRE Centre for Innovative Construction Materials (BRE CICM) team has attracted wide interest, and was recognised with the award of the Institution of Civil Engineers John Mitchell Medal in 2010 (with Dr Andrew Heath, Dr Chris Mundell, Claire Bailey and Professor Peter Walker).

Jean-Claude Morel was Director of Research (equivalent to professor) at École Nationale des Travaux Publics de l'État (ENTPE) in France, he is now Professor of Low Impact Buildings at Coventry University. He gained his PhD in 1996 from the University Joseph Fourier of Grenoble. In 1998, following an invitation from the architect Patrick Cohen, he initiated a series of research projects on drystone walls, at first with Boris Villemus, and later with Denis Garnier, Anne-Sophie Colas and then Eric Vincens. He has carried out major research work on low-impact building materials, including a series of testing campaigns on full-scale drystone retaining structures, on which he has published extensively.

Denis Garnier gained his PhD in soil mechanics in the prestigious École Nationale des Ponts et Chaussées (Paris, France), where he is now senior lecturer and teaches the course on continuum mechanics. He is also professor at ENTPE (member of the University of Lyons, France) where he teaches the course of Yield Design Analysis Theories. He started his research in the team of Jean Salençon with Patrick de Buhan. He has carried out major research work on rock engineering and the stability modelling of drystone structures, mainly based on homogenisation theories. He collaborates with Jean-Claude Morel of ENTPE, in 3D modelling and full-scale experiments of drystone masonries.

Chapter 1

Introduction

Structures built from carefully stacked stone, without any form of mortar, have been used throughout history for buildings and boundary structures, and for supporting terraces for agriculture, buildings and roads. They are an essential part of many monuments that have been designated as World Heritage Sites, such as Great Zimbabwe National Monument in Zimbabwe (1986) or the Historic Monuments of Ancient Nara in Japan (1998). Within France and the United Kingdom, a substantial proportion of retaining walls on roads are constructed of drystone: about one-sixth of road gravity retaining walls in France (Odent 2000) and about half of highway retaining walls in Great Britain (O'Reilly et al. 1999).

These structures have a long life. The drystone walls at Great Zimbabwe Monument were constructed in the eleventh century, and most road drystone retaining walls in France and the United Kingdom date from the nineteenth and the early twentieth centuries. However, the constituent materials are subject to weathering, especially to frost damage in colder climates, and so deteriorate over long periods. Poor repairs, especially pointing and grouting, can accelerate the deterioration by holding water within the structure, leading to accelerated weathering or even catastrophic collapse if significant pore pressures build up. Imposed loadings can be much higher than in the past due to increasing axle loads of modern vehicles, and even impact damage (Gupta and Lohani 1982).

Repair using appropriate methods or even full reconstruction may be needed. In the case of reconstruction, engineers need a design method that allows economical construction by avoiding overconservative design assumptions. Such methods need to be based on a proper understanding of drystone retaining wall behaviour, which in some respects can differ from the behaviour of conventional mortared masonry or mass concrete gravity retaining structures. These methods can allow efficient use of materials and resources, to produce structures that are both sensitive to the local environment and sustainable.

The stones of which the walls are made have been used predominantly as they have been found, often from clearing fields for cultivation. If the pieces of stone are defined by fractures on bedding or cleavage planes, they

Figure 1.1 A coursed masonry, limestone wall.

can be easily stacked to produce clear horizontal courses, with very little or no reshaping (Figure 1.1). If this is not the case, for example, with massive rocks such as granite, then shaping of the stones is usually carried out for the face of the wall (Figure 1.2). Then the construction process is no longer a matter of fitting flat pieces together in a plane, but of fitting lumps of rock together in a three-dimensional jigsaw, with few flat faces to make the process easier. In England, granite walling is particularly associated with the outcrops of the material forming the moorlands of south-west England and the Scilly Islands, but in France granite features strongly in part of the Cévennes, Alps and Britany.

Figure 1.2 A granite wall, France.

Blocks of granite could of course be shaped into rectilinear pieces that can stack easily, but only at the expense of a great deal of work. In some ancient civilisations, such as in South America, even hard rocks have been subject to considerable working so that the stones fit together with minimal gaps between them.

The nature of the stone is the principal factor that governs the way in which a drystone structure is built.

Because of their durability, stone structures are much more likely to survive for very long periods than structures made of earth, timber or other materials; only fired clay bricks are possibly more durable. Therefore, the proportion of surviving ancient structures that are made of stone does not indicate the proportion of structures made of stone in ancient times. The ability of stone to resist deterioration by moisture has led to it being favoured where it is in contact with the ground, which is a normal requirement of earth retaining structures, so the predominance of drystone technology in ancient earth retaining structures is not surprising. In any case, the simplest way to retain earth is with a massive material. Timber pile walls have been used in soft ground, particularly for quay walls, and timber has been used for bracing excavations, but using the weight of stone working with its rough surface has always been the simplest and most durable means of retaining soil – provided that a suitable stone is available locally.

1.1 USES OF DRYSTONE RETAINING WALLS

The most common reason for building earth retaining structures has been to provide level areas both in front of the wall (downhill) and behind the wall (uphill) for agriculture, some of the oldest examples supporting olive cultivation in Mediterranean countries. The underlying rock leads to natural slopes that are otherwise too steep for a good depth of soil to be stable, or to allow ease of movement and working, hence presenting the need. Weathered rock lying on or close to the surface, which may be removed to assist cultivation, or rock quarried from shallow benches, then provides the means. In such terrain, retaining structures also form terraces for buildings, or are part of the buildings themselves (Figure 1.3). In the steepest terrain, drystone structures were used to improve transport routes even for pack animals, but to this day they continue to be the most economical way to allow paths, roads, railways and even canals to be built along sloping ground in many parts of the world.

In hilly areas, retaining structures may originally have been built as boundary walls, with subsequent slope creep and hillwash progressively building up the ground level on the uphill side of the wall. As this filling occurs, the wall is raised through routine maintenance to continue to provide a barrier to livestock, and the weight of this parapet wall helps the wall below to retain the backfill.

Figure 1.3 Example of agricultural terrace with drystone retaining wall, Vialas, France.

1.2 CONSTRUCTION STYLES

This section will be developed further in Chapter 5. A drystone retaining wall may be constructed in the same way as a free-standing field wall and then backfilled. Most conveniently, the backfill will be placed in such a way that the current top of the wall where stones are being placed is kept at a comfortable working height above the backfill on which the wallers may stand.

In southern France, field walls are described as 'two-faced', and are something of a rarity, the predominant purpose of drystone wall construction being for earth retention, to support agricultural terraces or roads (Figures 1.3 and 1.4). There is also a strong tradition of drystone building,

Figure 1.4 Example of road with drystone retaining wall, Lozère, France.

Figure 1.5 A farmstead in the Cévennes, constructed of granite placed without mortar. Some recent pointing is visible in part of the wall.

both for simple cabanes (shepherds' huts) and for substantial farmsteads (Figure 1.5). For both purposes there is a strong motivation to achieve solidity, and walls are built with a fair back face for earth retention as well as for building construction.

In contrast, the British tradition of drystone wall construction comes from boundary walls, often made of stones cleared from the fields they enclose. These only need to resist wind load and impact by animals. They are traditionally made with two fair faces, and the gap between the backs of the stones forming each face contains 'fill', which should be well-packed small stones to allow good drainage while locking the main stones in place. 'Through stones' are then needed to tie the front and back faces together.

This British form of construction may then be adapted to provide a faster and less expensive means of building a retaining wall, in which the back face is simply omitted and replaced with a zone of rubble made up of the waste stone. This rubble is likely to have greater strength than the backfill, but could not produce a strong vertical wall of any great height, and this form of construction is more likely to be used with a substantial backwards batter (i.e., it leans back against the fill it retains), or in quite small structures. It is not considered further in this book, which is concerned with the construction of substantial and durable walls.

1.3 SUSTAINABILITY

Drystone structures have great strengths in terms of sustainability, but only if two rules are followed. The first is that they should be built using local materials, so that the energy used in transporting them to the construction site is kept to a minimum, and hence the stone block cannot be produced by a large industrial process. The second is that cement or lime is not used, as their production is responsible for a substantial proportion of the world's anthropogenic greenhouse gas emissions.

Provided these rules are followed, then the use of drystone masonry allows a technological leap in terms of sustainability. This will now be explored.

Before examining drystone retaining wall construction using the established methods for the quantification of construction sustainability, it will be useful to review the specific aspects which contribute to it. A qualitative approach is therefore followed first, and then a quantitative approach is given in Section 1.3.2.

1.3.1 Qualitative consideration of the sustainability of drystone retaining walls

1.3.1.1 Transportation of materials

The stone for the construction is often taken from very close to the site; in the case of repair or replacement, the old wall may provide most or all of the material required, depending on the stone's condition. The total transportation of materials, expressed in t · km, that is to say the total of mass times distance transported, will favour drystone compared to other modes of construction provided the stone has come from a local source. Otherwise, a more lightweight form of construction may be favourable in terms of transportation – the balance therefore depends on both the weight of the competing materials and the distance they have to be transported. This will obviously depend on the project and its geographic location – the stone might be considered 'local' if concrete and steel for an alternative would have to be transported much further, and a careful comparison may be needed. To give an indication, the study published by Morel et al. (2001), which compared stone masonry with earth mortar to build houses, showed that if the supply of stone was within 50 km of the construction site, the transport of material remained below that of an equivalent concrete construction.

Life cycle assessment (LCA) attempts to provide a simple measure to support simple decision making, whereas it is important to assess transport separately, because of its impact on the road network, on road accidents and on air quality. Because drystone is so important in hilly and mountainous regions, where the roads may be narrow and not suited to very heavy traffic, it is important that the transportation is assessed properly.

1.3.1.2 Socioeconomic issues

There is very little scope for the 'industrialisation' of drystone construction, because each stone must be placed by hand. This requirement for *skilled* manual labour is very important – there is much more need for skill than in any other system of construction of retaining walls. This job is very rewarding for workers, because it calls on a range of intellectual abilities that go beyond the mere academic, in addition to physical skill and strength, thus giving the builder the opportunity to develop a range of skills and abilities that are very conducive to mental and physical health and fitness. Many companies and organisations see volunteers of all ages aiming to become drystone wallers; the manual trades and crafts in general, and particularly in construction, are less attractive to young people. The high social intensity of the work encourages wider community stakeholders to be very positive in proposing new drystone retaining walls.

The corollary is that the cost of dry stone construction is contained mainly in the workforce (and not in the material); thus in societies in which labour is expensive relative to other energy sources, it would greatly disadvantage drystone construction. Indeed, in Europe and especially in France, the price of non-animal energy (electricity, gas and oil) is 100–200 times less than human energy (Marcom 2002; Marcom et al. 2009; Rigassi and Seruzier 2002).

In conclusion, should the costs of non-human energy increase, the difference in costs between human power and other energy sources decreases, which would reduce the cost of building in drystone, and make this type of building less expensive and more profitable. A substantial part of this equation comes from the way in which taxes are extracted from the economy. The cost of energy is in reality anything but a free market, being heavily distorted by both state taxes and the manoeuvrings of the energy producers, especially when they too are states. In the longer term, states are keen to use taxation on energy costs to reduce the fruitless generation of entropy and hence impact on the environment, for this increases the income to the politicians, while the increasing efficiency of production reduces the requirement for labour to carry out mundane tasks. In the long term, it should be expected that in every country the workforce becomes more highly skilled, and the incentives for using skilled labour to reduce environmental impact will only increase.

1.3.1.3 Durability

Drystone retaining walls often have remarkable durability, but this depends on the geology of the site, and hence the nature of the stone, and on the quality of maintenance. It also depends on the design and on the quality of construction.

Compared to current design criteria, historically some walls were undersized, especially retaining walls on farmland, for example, which were most probably for the most part developed through a process of trial and error to produce very efficient designs for a very limited range of materials and wall heights. This strategy was used because it was more reasonable for farmers to accept occasional failures in extreme circumstances than to systematically build oversized walls, as is the practice of modern engineering.

1.3.1.4 Reusing the materials

The use of rubble stone without mortar allows reuse directly for 'new' building, but a proportion of the stone will be lost through frost-shattering, for example. This practice has continued for thousands of years, and is still practiced. Reuse of materials allows a high degree of sustainability, even if the resource is not renewable in the absolute sense. The potential use of reusable material does not affect biodiversity, and does not require the use of any agricultural area, unlike intensive short rotation forestry, for example. Following Bruno Durand (ABPS, drystone retaining wall contractor, personal communication, 2009), it is safe to assume that 30% of the stones can generally be reused.

1.3.2 How to measure the sustainability of drystone walls

General tools currently used to measure sustainability are still in development, but they have been developed for industrial materials and structures, as for example, LCA. Following the discussion above, these tools, despite their complexity, might not be well suited to drystone retaining walls, and in any case, the necessary databases for their use are not yet available for drystone. However, it is always interesting to test these methods to be aware of their limits, which is why the following case study is examined.

1.3.2.1 Life cycle assessment

The sustainability analysis is carried out through the case study of a recent construction in the district of Felletin, Creuse, France (Colas et al. 2014a). Professional drystone masonry wallers were appointed to build a drystone retaining wall to support a local road in 2012. The wall is 3 m high and 50 m long, and made of recovered granite blocks (Figure 1.6). Works were completed in 6 weeks and required 12 masons.

The LCA was carried out following standard practice, and referring to NF EN ISO 14040 and NF EN ISO 14044. It consists in evaluating the resource usage and environmental impacts of a product or a service. The four steps identified in the standards are detailed below.

1.3.2.1.1 Goal and scope definition

LCA starts with the definition of the context and research topic of the study. Here, the functional unit is a civil engineering construction retaining

(a)

(b)

Figure 1.6 Case study of a drystone retaining wall built in 2012 in Felletin, France. (a) The wall on completion. (b) The wall being loaded by a lorry and measurement of the deflection.

a backfill supporting a vehicular road, for a lifespan of 100 years. In the first approach, the boundaries of the system are limited to the construction of the retaining structure: production of materials, transport of materials and equipment and the construction stage. The study concentrates on the drystone wall, the comparison with a concrete solution being discussed as a perspective.

1.3.2.1.2 Life cycle inventory

The data related to the construction have been collected with the support of the drystone wallers who built the wall; the main data are presented in Table 1.1. This enables the inventory of flows involved in the construction to be prepared.

Table 1.1 Life cycle inventory from the construction of the drystone retaining wall in Felletin

Stage of building process	Type of materials or equipment	Quantity
Material production	Granite	110 m³
Transport	Materials	414 t · km
	Heavy equipment	3330 t · km
Construction stage	Diesel for equipment	960 L
	Personnel transport	15,750 km

Source: Colas, A. S. et al., Holistic approach of a new masonry arch bridge on a Cevennes road. In *9th International Masonry Conference 2014*, Guimarães, Portugal, July 7–9, 2014.

1.3.2.1.3 Life cycle impact assessment

The flows of the life cycle inventory are used as an input to calculate the environmental impacts of the construction of the drystone wall. Ten impact assessment categories have been chosen for this study, referring to the methodology developed at the Center of Environmental Science of Leiden University in 2001 (Frischknecht et al. 2007). The evaluation has been performed using Simapro 7.3.3 software and the Ecoinvent database. It enables the identification of the most important processes for environmental impacts. Two hypotheses have been explored on the origin of the stones composing the wall: recovered stones, in accordance with the real project, and stones from a local quarry, to include general cases.

1.3.2.1.4 Interpretation

In the case of construction with entirely recovered stone, the contribution of the material production stage is very low (Figure 1.7a), as the sole inputs are the infill material and the amortisation of heavy equipment. In the case of drystone reconstruction, this assumption is partially fulfilled, as a part of the stones can be easily reused. When the stone comes from a local quarry, the contribution of the material production increases, but still proves to be roughly equivalent to the transport and constructions stages.

1.3.2.1.5 Economic issues

One of the main reasons behind the reduction in use of drystone techniques in modern construction is common belief about its cost. The drystone wall in Felletin cost €2500 per linear metre. Economic data on civil engineering works are very difficult to find and compare considering the specificity of each project. However, this can be compared to figures given in O'Reilly et al. (1999), where the cost of the replacement of all drystone retaining walls in Great Britain was estimated at €1400 per linear metre in 1999. This can also be compared to expert advice (Alava et al. 2009) estimating the cost of a 3 m high concrete wall as €3200 per linear metre. Thus, the cost of drystone walls stands in the same value range.

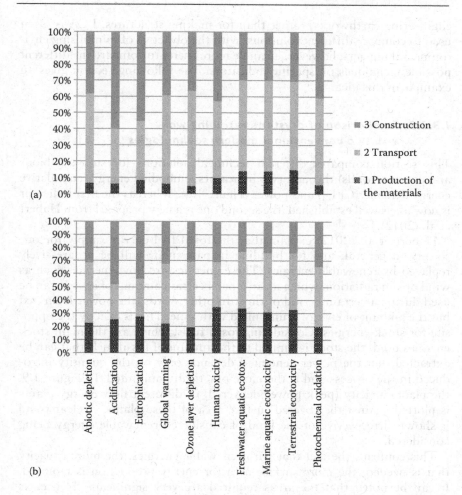

Figure 1.7 Identification of the contribution of material production, transport and construction stages in the environmental impacts of the construction of the dry-stone retaining wall in Felletin: recovered stones (a) and stones from a local quarry (b). (From Colas, A. S. et al., Holistic approach of a new masonry arch bridge on a Cevennes road. In *9th International Masonry Conference 2014*, Guimarães, Portugal, July 7–9, 2014.)

1.3.2.1.6 Conclusion

Broader issues include the integration of maintenance over 100 years to analyse the structure over its whole life cycle. Recent studies by Colas et al. (2014b) have proved the importance of considering the life cycle of the structure to choose the most appropriate solution. The weakness of LCA is that the chosen scales may not be appropriate for this type of construction with earth materials, which are predominantly used in bulk for civil engineering earthworks, rather than for making structures. LCA is often used to compare different solutions, with the objective of minimising environmental impacts; however, it can be more useful to compare the different possible techniques on specific indicators. The following section gives an example of this idea.

1.3.2.2 Comparison of drystone retaining wall with two conventional modern technologies

This section compares different technical solutions (drystone, gabions and concrete walls) through two indicators: embodied energy (cumulative energy demand [CED]) and power (Figure 1.8). Whereas the first indicator is nowadays well established, the second one is a new proposal from Habert et al. (2012).

Habert et al. (2012) explain that the fossil fuel energy supply for our society in general, and for building in particular, will be progressively replaced by renewable energies. These energies are flow energies such as wind or sun radiation, which means that a certain amount of energy can be used during a certain period of time. In other words, the power is limited but the amount of energy is unlimited with time. This is exactly the opposite for stock energies. Considering fossil fuels, which are the main stock energies used, the stock is limited to the amount of fossil fuels that can be extracted, but the power generated depends only on the quantity introduced in the process and is therefore practically unlimited. In Figure 1.9, the plant capacity (peak power delivered) of different flow energy plants is plotted against the covered surface area of these plants. A clear trend is shown, irrespective of the kind of considered renewable energy being considered.

This confirms the fact that with renewable energies, the more capacity that is needed, the more surface area for energy production is required. It can be noted that the areas required are very significant. Hence, as the reduction of the use of fossil fuel will probably be associated with an increasing use of renewable energies (Omer 2008), the principal characteristic of future societies will be power restriction, rather than energy depletion. Figure 1.9 illustrates that the maximum power required for product manufacturing is a much more critical problem than considering the average continuous power needed every day to supply people's needs.

1.3.2.2.1 Power calculation method

As different materials are involved in the construction and the service life of buildings, the specific power for each material is first needed. A detailed study of the different processes leading to the production of a material must be performed to identify which process needs the highest power. It is this power that will be considered as it is the one that needs the largest surface

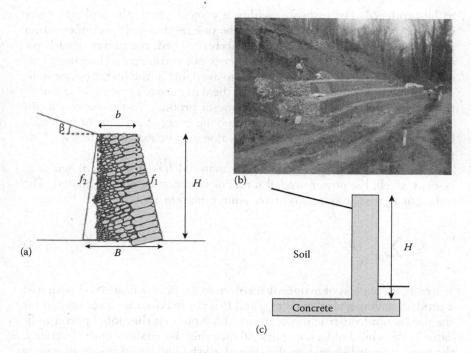

Figure 1.8 Comparison among three technologies: (a) drystone retaining wall, (b) gabion, (c) reinforced concrete (cantilever wall).

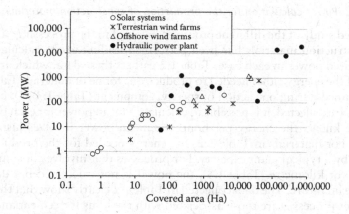

Figure 1.9 Relationship between the covered surface for energy production and the resulting power capacity generated for various renewable power plants. (From Habert, G. et al., *Ecological Indicators* 23: 109–115, 2012.)

to be produced. For instance, within a cement plant, the highest power involved in the cement production is the one needed for the clinkerisation process of the clay–calcareous rock mixture. Then, the power needed for other processes such as grinding clinker is not considered. The power calculation can be done directly if data are available in the literature, or indirectly if one has access for instance to the daily capacity of a plant, and to the energy needed to produce one tonne of product. In this case, a mean daily power can be calculated by multiplying energy per tonne of materials by the tonnes produced per day. This value may be expressed in $J \times s^{-1}$, or power (W).

Once the required power for each material has been calculated (e.g. cement, steel), the power needed for a construction can be calculated. The indicator P_g for the system is given using Equation 1.1.

$$P_g = \sum_i \frac{m_i}{m_{tot}} P_i \tag{1.1}$$

where m_i is the mass of material i involved in the production/construction of a product having a total mass m_{tot} and P_i is the maximum power needed for the production/construction of material i. Note that the global power indicator P_g does not hold a true physical meaning. By analogy with electricity, the power needed should be the sum of all the individual power processes (ΣP). In this new indicator, the individual power processes are weighted by their mass contribution to the global product. As a result, for two products A and B, if $P_g A$ is higher than $P_g B$, it implies that product B involves less amount of materials depending on high power processes than product A.

1.3.2.2.2 Power calculation for the production of construction materials

A detailed study of the different processes involved in the processing of different construction materials has been performed leading to the calculation of the required power in each case. Table 1.2 gathers these data, which are composed of the energy either needed to produce one tonne of material (Table 1.2) or consumed during one action of specific equipment (Table 1.3). Once these energies are collected, it is possible to calculate a mean power for each process.

If one knows the energy per hour, the mean power involves just a unit change. For materials in Table 1.2, the energy needed for the production is divided by a typical plant capacity. For processes that involve cubic metre of concrete or kilometre (Table 1.3), the power is not calculated but deduced from the literature. The data collected in Table 1.2 clearly show that the highest power processes are those associated with the kilns for cement and steel. Table 1.4 gathers the maximum power needed in the production process of each material used for the building construction. For aggregate production, the highest power of the processes presented in Table 1.3 has been used.

Table 1.2 Mean power of different equipment to produce one ton of material

	Energy (MJ/t)	Daily production (t)	Mean power (MW)	Reference
European Cement kiln	3600	4000	167	Bastier (2000) and JRC (2000)
Blast furnace	12 726	4107	605	Classen et al. (2007)

Source: Habert, G. et al., Ecological Indicators 23: 109–115, 2012; Martaud, T., Evaluation environnementale de la production de granulats naturels en exploitation de carrière: indicateurs, modèles et outils. Ph.D. thesis. Orléans University, France, 212 pp., 2008.

Table 1.3 Mean power consumed during one action of specific equipment

	Unit	Energy (MJ/Unit)	Mean Power (MW)	Reference
Loader (Caterpillar 950 F)	hour	657.4	18.3×10^{-2}	Martaud (2008)
Dragueline	hour	484.4	13.5×10^{-2}	Martaud (2008)
Jaw crusher (38–156 m³/h capacity)	hour	475.5	13.2×10^{-2}	Martaud (2008)
Spring cone crusher (PYD1750)	hour	576.5	16.0×10^{-2}	Martaud (2008)
Diesel generator (10 kVA)	hour	85.9	2.4×10^{-2}	Kawai et al. (2005)
Truck (20 t)	t · km	1.03	2.1×10^{-2}	Kawai et al. (2005)
Concrete pump truck (40–45 m³/h)	m³	6.19	6.8×10^{-2}	Kawai et al. (2005)
Ready mix plant	m³	99	10.0×10^{-2}	Chen (2001)
Crawler excavator (0.6 m³)	hour	260	7.2×10^{-2}	FNTP (2010)

Source: Habert, G. et al., Ecological Indicators 23: 109–115, 2012; Martaud, T., Evaluation environnementale de la production de granulats naturels en exploitation de carrière: indicateurs, modèles et outils. Ph.D. thesis. Orléans University, France, 212 pp., 2008.

Table 1.4 Energy and power for the production of materials used in the case study

	Fossil cumulative energy demand (MJ/t)	Maximum power needed in production process (MW)
Fired brick	3.0×10^3	1
Cement	4.7×10^3	167
Steel	24.4×10^3	605
Aggregates	1.0×10^2	18.3×10^{-2a}
General timber	8.5×10^3	18.3×10^{-2}
Stone	3.0×10^2	18.3×10^{-2a}

Source: Habert, G. et al., Ecological Indicators 23: 109–115, 2012.

[a] It is considered that for general timber, power of materials will not be different from those used for aggregates or stones.

The two types of retaining walls that have been studied to compare with a drystone retaining wall are a gabion retaining wall and a reinforced concrete (cantilever) wall. The three structures are presented in Figure 1.8.

The drystone retaining wall considered is 3 m high, being the most common height found in the European heritage. The friction angle of the soil is assumed to be 30° (for a sandy soil), the natural slope of the backfill is set at 20° and the batter f_1 is fixed at 20% and the internal batter f_2 at 0%. The design procedure follows the principles of Chapter 4 and is given in the charts in the Appendix. Then, by estimating the void ratio within the stone masonry at 25% (see Chapter 3) and the stone density at 2650 kg/m³, the stone volume for 1 m wall length is equal to 3.1 m³, or 8228 kg. There is of course a possibility that some of this stone is already on site and is reused.

The gabions being considered (Figure 1.8b) are electro-welded wire mesh gabions, with square or rectangular mesh, which gives a very good rigidity. The fill materials for the gabions are large aggregate materials with the highest possible density and frost resistance, but crushed concrete can also be employed. For a better basket fill, the greatest stone dimension is generally limited to 250 mm. The baskets are often installed in their final place on site, then filled and closed, which provides more efficiency than if they were assembled in a factory. Indeed, in this way no crane is required to assemble them on site. The quantity of stones required to build a retaining wall with gabions is similar to that required for a drystone retaining wall. The information related to the steel quantity involved in gabion technology can be found in Table 1.5.

For the cantilever walls, the stability is ensured by the rigidity of the wall itself but also by friction between the wall and the foundation (if the bearing capacity of the soil foundation is correct). The wall is then as thin as possible with steel reinforcement to resist tensile stress due to the bending moment. This type of wall is very common and can be calculated using the classical strength of materials theory.

1.3.2.2.3 Energy and power associated with retaining walls

The materials used to build the three types of walls are indicated in Table 1.5 and the equipment involved in the process of construction is described

Table 1.5 Materials used to build the three different walls 3 m high

	Concrete	Gabion	Drystone
Excavation (m³)	7	7	7
Cement (kg)	672	0	0
Steel (kg)	240	40	0
Aggregates/stones (kg)	4680	8228	5759

Source: Alava, C. et al., Murs de soutènement: Comparaison environnementale et financière de différentes technologies. Projet d'Option, École Centrale de Lyon, 2009.

in Table 1.6. The energy and power data used in this process are presented in Tables 1.2 through 1.4. Table 1.7 allows comparison of the power indicator (P_g) to the embodied energy indicator (CED) and also to the sum of the power (ΣP) used during the construction. In Table 1.7, P_g is the most sensitive indicator distinguishing the three technologies. Using ΣP, the values obtained for gabion and concrete walls are rather close. If the concrete value is set to 100%, then gabion ΣP is equal to 78% instead of 53% if the calculation is done with P_g. If CED is used, then the values obtained for gabion and dry stone walls are closer than with P_g (Table 1.7). When the embodied energy of the concrete wall is set to 100%, the gabion and dry stone walls have an embodied energy of respectively 27% and 8% of the concrete wall

Table 1.6 Equipment required for the construction of retaining walls 3 m high, per linear metre

	Unit	Concrete	Gabion	Drystone
Excavation works				
Crawler excavator	hour	0.33	0.33	0.33
Materials transport				
Concrete	km	15		
Cement (by 9-tonne truck)	km	150		
Steel (by 28-tonne truck)	km	500	500	
Aggregates (by 28-tonne truck)	km	25	25	25
Construction works				
Man-made	hour			30
Crawler excavator	hour		2	
Diesel generator (10 kVA)	hour		2	
Ready mix plant	m³	2.5		
Agitator truck (4.5 m³)	m³ · km	37.5		
Concrete pump truck (40–45 m³/h)	m³	2.5		

Source: Alava, C. et al., Murs de soutènement: Comparaison environnementale et financière de différentes technologies. Projet d'Option, École Centrale de Lyon, 2009.

Table 1.7 Embodied energy (total CED), power sum (ΣP) and proposed power indicator (P_g) calculated for the three different walls, with the concrete value set at 100%

	Concrete	Gabion	Drystone
Energy indicator, total CED (MJ)	10.26×10^3 (100%)	28.02×10^2 (27%)	8.10×10^2 (8%)
Power sum, ΣP (MW)	7.72×10^2 (100%)	6.05×10^2 (78%)	3.19×10^{-1} (0.04%)
Power indicator, P_g (MW)	3.97×10^2 (100%)	2.10×10^2 (53%)	1.72×10^{-1} (0.04%)

Source: Habert, G. et al., Ecological Indicators 23: 109–115, 2012.

embodied energy. When a similar comparison is performed on the basis of P_g, the relative values are equal to 53% and 0.04% for gabion and dry stone wall, respectively. Therefore, the indicator P_g enhances the differences between a structure in which no power intensive materials are involved (drystone wall) and a structure in which a power-intensive material such as steel is used, even in small quantities (gabion walls). Therefore, this study of three very different structures, for which intuitively there are substantial differences, shows that P_g is able to make the clearest distinctions compared to the other indicators.

1.4 SUMMARY

This book does not address the sustainability of drystone retaining walls in detail, but it is important to know that drystone is the best material in terms of sustainability, provided that certain rules are respected, as has been described in this chapter. Chapter 2 gives an introduction to the behaviour of simple earth retaining structures. This is fundamental because the load applied to the drystone retaining wall by the backfill soil will determine the initial design of the retaining wall. Chapter 3 then deals with those aspects of the behaviour of drystone retaining structures that are obviously very specific to this type of material: no tensile strength and a dependence on the arrangement of the stones. Chapter 4 provides the theoretical framework to model the drystone retaining wall's mechanical stability by yield design. Chapter 5 provides the rules for the construction of the wall; respecting these rules will ensure that the theoretical assumptions of the model are realised in the actual wall, but will also ensure that the wall has the ductility and resilience which is a particular characteristic of this form of construction. Chapter 6 then gives guidance regarding the assessment of existing structures, to be able to make good engineering decisions about their future.

Chapter 2

An introduction to the behaviour of simple earth retaining structures

Drystone retaining walls are usually thought of as gravity structures – that is, they resist the pressure of the earth behind them solely by their self-weight. This is not quite accurate, but consideration of the simple gravity structure allows some basic concepts to be covered.

For the wall to stay in place, the forces acting on it must be in equilibrium. The effect of forces acting out of alignment with each other results in a moment or couple, which would cause a rotation. We therefore require there to be an equilibrium of forces, as well as an equilibrium of the resulting moments – summarised as force and moment equilibrium. This equilibrium needs to exist for the wall as a whole, and for individual parts of the wall that may move relative to each other. The terms used to describe different parts of the wall, and the soil around it, are shown in Figure 2.1.

2.1 THE BEHAVIOUR OF SOIL

Even if the structure is fully in equilibrium, the forces acting on it cause stresses within it – these may be compressive (compression), tensile (tension) or shearing (shear); see Box 2.1. These stresses can be considered at the level of the wall as a whole, or at the level of its individual components.

Compression of a material will cause it to shorten in the direction of the compression, and if the compression is not equal in every direction, there will be consequent shearing stresses that result in the material changing shape – it becomes wider as well as shorter (Figure 2.3a). The rock of which drystone retaining walls are made is usually strong and stiff in compression – it takes a large load to break it, and it does not deform very much at all before it breaks. The compression of a complete wall will usually also be slight.

Tension causes a material to lengthen, and this will usually be accompanied by a thinning (Figure 2.3b). The tensile strength of stone and concrete may be quite high, for example, 2–5 MPa for concrete, compared with a compressive strength of 20–40 MPa. To get a sense of this, consider that a solid concrete block 100 m high would generate a pressure of only 2.5 MPa

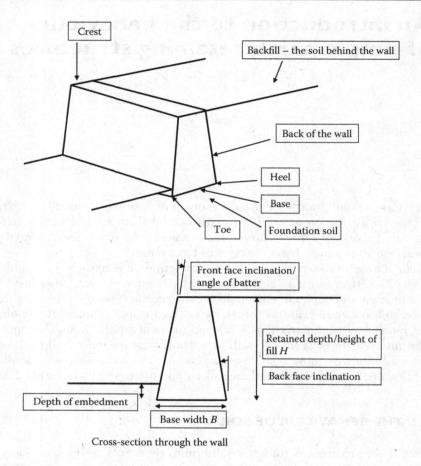

Figure 2.1 Terminology.

of compression on its base. It is a strong material, suited to making the frames of buildings and bridges, but to make a whole retaining wall out of it would be very wasteful. For this reason, earth retaining structures made of concrete, usually exploit the tensile strength of steel reinforcement in relatively thin sections of concrete using the combination of steel and concrete to resist bending moments and the weight of the soil itself to ensure stability (Figure 2.4). For comparison, a strong limestone might have a compressive strength of 50 MPa, and a tensile strength of 5 MPa. However, the tensile strength of a material, as of a chain, is only that of its weakest link. In cemented masonry, this is usually that of the mortar, or the bond of the mortar with the stone or brick. In drystone masonry, there is no mortar,

BOX 2.1 STRESSES IN SOIL

Soil is made up of individual particles of rock and clay minerals, and the forces that are carried by the soil are transmitted from particle to particle, where they touch. In most soils, these individual particles are quite strong, but they are just resting on each other and can be moved. So a heavy enough load might push the particles into a new arrangement in which they are closer together – this is called compression. On the other hand, if this heavy load is concentrated, then particles can rearrange, not just move closer together, so that the shape of the body of soil changes – this is shearing, and it can also happen if particles are pushed to one side directly. Shearing can result in much bigger movements than pure compression. Compression can lead to settlement of foundations, but shearing can result in an entire hillside sliding down to the floor of a valley. If you look closely enough at the soil, you always see individual particles moving relative to each other, but step back and what you often see is that the rearrangement of particles that produces the shearing movement is concentrated on a clearly defined surface, called the shear band.

Advanced computer analysis can model the individual particles and the forces at the points of contact between them for relatively small problems, but for nearly all real problems it is necessary to take a step back from the individual particles and take an overview. This means looking at what is happening on a notional plane within the soil, and considering the total force per unit area on that plane. This is called the stress. We can consider the components of force acting at right angles to the plane, and the components acting along the plane – these give the normal stress σ and the shear stress τ. The normal stress in soil is taken to be positive when it is compressive (a pressure), and there are only two directions in which the normal stress can act, as it must by definition be at right angles to the plane, giving tension (negative stress) or compression (positive stress). The direction of the shear stress, however, is within the plane (Figure 2.2), but can be in any direction within the plane.

Given that the plane that we are looking at can be in any orientation, this consideration of stresses in three dimensions can become very complicated and confusing. Fortunately, the situation for retaining walls can be simplified considerably because we can provide a good representation for many purposes by looking at a two-dimensional cross-section through the wall, as shown, for example, in Figure 2.5.

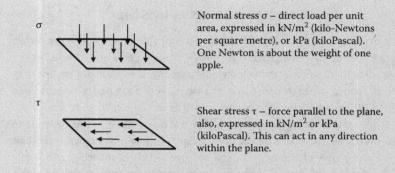

Normal stress σ – direct load per unit area, expressed in kN/m² (kilo-Newtons per square metre), or kPa (kiloPascal). One Newton is about the weight of one apple.

Shear stress τ – force parallel to the plane, also, expressed in kN/m² or kPa (kiloPascal). This can act in any direction within the plane.

Figure 2.2 **Normal and shear stress on a plane within the soil.**

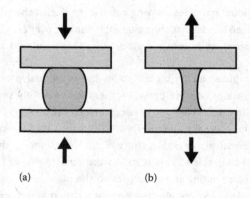

(a)   (b)

Figure 2.3 **Effects of (a) compression and (b) tension.**

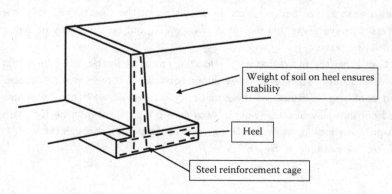

Weight of soil on heel ensures stability

Heel

Steel reinforcement cage

Figure 2.4 **A reinforced concrete cantilever retaining wall.**

and the tensile strength of the material as a whole might be nil if the rocks can come apart in the direction of the tension. However, a drystone wall can have considerable tensile strength along the length of the wall if the stones overlap each other (have good bonding), as the weight of the wall sitting on top of them prevents them from sliding past each other. This is described in detail in Chapter 3.

Shear stress causes a material to change shape and is particularly significant in soil, being the dominant mode of failure; the tensile strength is almost zero, and compression of the soil will usually lead to a failure in shear rather than crushing, because it is very rare for the compression to be uniform in all directions. Shear deformation is often concentrated on a clearly defined shear surface, which may be the junction between two different materials. Within a drystone structure, the shear deformation is predominantly a result of one stone sliding some distance across the top of another stone, though large shear forces can be sustained before significant movement takes place.

Provided that the compressive, tensile and shear stresses in a structure are not close to or exceeding the corresponding strengths, the resulting deformations will usually be very small. The compressibility of the soil the structure is sitting on is much more likely to result in significant movement.

2.2 THE FUNDAMENTAL REQUIREMENTS OF A GRAVITY RETAINING WALL

Figure 2.5 shows cross-sections of a simple gravity retaining wall, with the retained soil to the right. The soil rests against the back of the wall, exerting a pressure on it as represented in Figure 2.5a. This pressure arises from the weight of the soil, and so increases with depth behind the wall. Unless the back of the wall is exceptionally smooth, there will be significant friction on the back of the wall that will help to support the weight of the soil and so reduce the horizontal pressure on the wall. This friction, together with the weight of the wall itself, exerts pressure on the ground beneath, which holds the wall in vertical equilibrium. The wall will sink a little as it is built, compressing the soil as the pressure on the base increases. Provided the ground is not overloaded, no failure will occur, and vertical equilibrium will be maintained.

The pressure on the back of the wall tends to move the wall forwards, by sliding on its base (Figure 2.6a). This is resisted principally by friction between the base of the wall and the soil it is sitting on. Most walls have their bases at an embedment depth of between 0.5 m and 1 m below the surface of the ground in front of them. This comes from removing topsoil and other weaker material before the wall is built, with greater depths being used to ensure that the ground the wall is sitting on is strong enough to support its weight and to ensure that the wall is not resting on ground

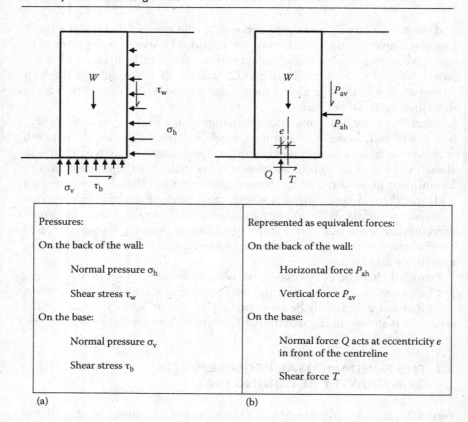

Figure 2.5 (a) Pressures and (b) forces on a simple earth retaining structure.

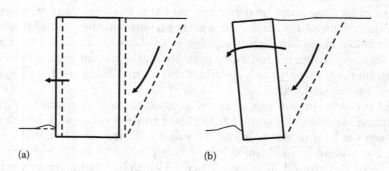

Figure 2.6 Failure modes. (a) Sliding. (b) Overturning.

that might freeze and expand. This means that there is usually a shallow depth of soil in front of the wall that must be pushed forwards if the wall is to move. This soil can sometimes be removed, for example, when digging trenches for services, and may not be there during the construction, so it is normal to ignore this 'passive resistance' in the design. Provided sufficient friction can be mobilised on the base of the wall, then horizontal equilibrium will be maintained, and the wall will not slide forwards.

The wall must also be in equilibrium against rotation. If the wall is tall and thin, it can be seen that the pressure of the earth on the back of the wall will cause it to rotate about the toe – to fall over forwards (Figure 2.6b). Before this happens, most of the weight of the wall will be resting on a thin strip of soil just behind the toe, and this is likely to give way first, precipitating a failure significantly more easily than might be expected. It should be noted that friction acting downwards on the back of the wall helps to prevent this overturning failure, so a construction that presents a rough surface to the backfill will be more stable.

2.3 EARTH PRESSURE CALCULATIONS

Dry soil tipped out onto a level surface will form a pile with sides sloping at the angle of friction of the soil, which is why it is also known as the 'angle of repose'. To make the side slopes any steeper requires something for the soil to rest against which must be capable of supporting a minimal pressure – this is the function of a retaining wall. All the soil between the angle of repose and vertical must be held in place by the retaining wall.

If the backfill is compacted in place, then part of the pressure applied during the compaction will be applied to the wall – but the wall only needs to move a very small amount for the pressures to reduce towards the minimum values. Soil is usually compacted when it is placed to ensure that it is uniformly dense, and does not settle unduly over the course of time, either due to its own weight or due to applied loads. However the backfill is placed, it will rub down against the back of the wall, and if the wall is at all rough it will exert a downwards force as well as a horizontal force.

The simplest assessment of earth pressure is for a wall with a perfectly smooth back on which no friction is exerted. In this case the horizontal pressure is simply related to the vertical pressure in a way that is governed by the shear strength of the soil (see Box 2.2). The minimum horizontal pressure that must be supported by the retaining wall comes from the geometry of the Mohr circle:

$$\sigma_h' = \sigma_v' \, \frac{1 - \sin(\varphi')}{1 + \sin(\varphi')} \tag{2.1}$$

BOX 2.2   THE STRENGTH OF COHESIONLESS SOIL

It is easy to represent the strength of the soil by a coefficient of friction – this accords with the observation that the greater the direct pressure on a surface, the more difficult it is to cause shearing on that surface. If the surface is a plane parallel to a slope, then the coefficient of friction is equal to the tangent of the angle of friction if the slope is as steep as it can be. This leads geotechnical engineers to use the 'angle of friction', φ', as the coefficient of friction, rather than a simple coefficient μ.

We can then consider the relationship between this frictional strength and the stresses within the soil. We might start by asking what is the maximum shear stress, but in fact, because the strength is frictional, we need to know the maximum ratio of shear stress to normal stress:

$$\tau_{max} = \sigma' \tan (\varphi') \tag{B2.2.1}$$

So because τ and σ' depend on the orientation of the plane, the strength of the soil will be fully used if

$$\tau_{max}/\sigma' = \tan (\varphi') \tag{B2.2.2}$$

We therefore need to look at different orientations of the plane to find the orientation that gives us this value, but we only need to rotate the plane about a direction which is at right angles to the cross-section, that is, along the line of the wall (Figure 2.7).

We can then represent the stresses acting on a plane on a two dimensional plot of shear stress and normal stress, as shown in Figure 2.8. If the stresses are plotted as the plane is rotated through 180°, they trace a circle, known as the Mohr circle. On the same plot, we can draw lines showing the limiting strength of the soil, $\tau_{max} = \sigma' \tan (\varphi')$. Note that these make an angle of φ' with the horizontal (σ') axis. If the stresses are such that the circle touches these lines, then we are interested in the point where it does so, because this will be the state of stress on planes on the point of shear failure, and we can find out the orientation of those planes. Because of the geometry of the circle, the orientation of the planes can be drawn within the circle as shown in Figure 2.8. This plot is very useful, as it can show us the stresses in soil behind a retaining wall if the strength of the soil is being exploited fully. To express this another way, it can show us the horizontal stress which must be provided to prevent the soil from going into shearing failure.

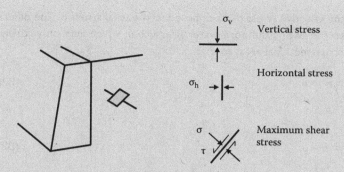

Figure 2.7 Rotation of a plane within the soil.

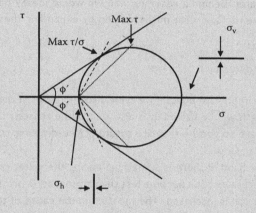

Figure 2.8 Mohr circle of stresses at a point.

The angle of friction can vary from 21° or so for clay soils, through 32°–35° for sands, to more than 40° for gravels. The larger particles tend to have larger friction angles because they are rougher and more angular. Most natural soils, and most soil used as fill behind retaining walls, contain a range of particle sizes. The smaller sized particles make it difficult for water to flow through the soil, and in silts and clays the space between the particles can be filled with water. If the water is prevented from flowing out of the soil, then the pressure in the water will increase with depth. This reduces the forces at the points of contact between soil particles, by the buoyant support of some of the weight of the particles. As these contact forces give rise to the frictional strength of the soil, the presence of positive water pressures reduces the frictional strength of the soil. We express this by talking about the effective stress σ', as opposed to the total stress σ. The difference between the two is the pore water pressure u, which acts on virtually the entire cross-sectional area, so that:

$$\sigma = \sigma' + u, \tag{B2.2.3}$$

or

$$\sigma' = \sigma - u \tag{B2.2.4}$$

The total stress is controlled mostly by the self-weight of the soil, and any loads imposed on it, whereas the water pressure is controlled mostly by the depth of water. Behind a retaining wall we would ideally have no depth of water, but we can see what might happen by expanding the equation for shear strength:

$$\tau_{max} = (\sigma - u) \tan (\varphi') \tag{B2.2.5}$$

This is a very important effect – for typical soil densities, if the soil is full of water its strength will be halved. It is also one of the reasons that drystone retaining walls are so good – they are naturally free-draining, so pore water pressures are very low.

On the other hand, if there is something pulling the water out of the soil then the pore pressures can become less than atmospheric pressure, and the strength of the soil is increased. The most common cause of this is vegetation. A similar effect occurs if the soil is trying to expand, as will happen if a lump of soil is dug out of the ground and no longer has pressure acting on it. This soil will be trying to suck air into it as it expands, and so the water pressures will become negative. However, if the particle sizes are small, then the effect of water surface tension around the edge of the soil is to maintain a big pressure difference between the air and the pore water, and the lump of soil remains intact and strong. This is how sandcastles work – the finer the sand, the stronger the effect. If the particles are as small as clay, the effect can become strong enough to make building bricks which do not need to be fired in a kiln. This is important for older retaining walls that were often backfilled with clayey soil – it means that the backfill can stand up for some time without exerting any pressure on the wall at all, and can stand up for a period without a wall in place. This can lead to an inappropriate sense of security – unsupported cohesive soil could fail quickly with little warning, especially if the ground becomes wetter following heavy rain.

From this we get

$$\frac{\sigma'_h}{\sigma'_v} = K_a = \frac{1 - \sin(\varphi')}{1 + \sin(\varphi')} \tag{2.2}$$

where K_a is known as the coefficient of active earth pressure. This presents a very simple picture of the earth pressure on the back of a retaining wall, as shown in Figure 2.9. The vertical stress at a depth d is just d multiplied by the unit weight of soil γ, typically in the range 17–20 kN/m³. The horizontal stress is then just K_a times the vertical stress. We can then work out the total force required to resist this pressure distribution. The horizontal stress varies from zero at the ground surface to $K_a\gamma H$ at the bottom of the structure, so the average pressure is $K_a\gamma H/2$, which acting over the full height of the wall gives the active pressure force $P_a = K_a\gamma H^2/2$.

Another way to reach this result is to consider not the stresses in the soil, but the mechanism by which it might fail (Figure 2.10), which may

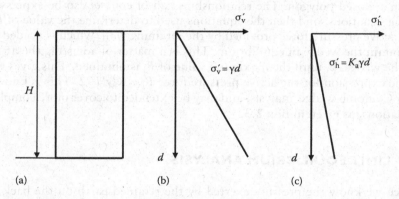

Figure 2.9 Simple earth pressure. (a) Wall cross-section. (b) Vertical stress σ'_v. (c) Horizontal stress σ'_h.

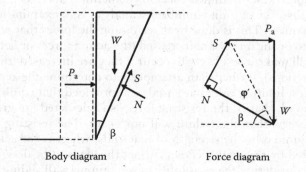

Figure 2.10 Wedge analysis.

be described as limit equilibrium analysis. We can speculate that as the limiting condition is reached, a wedge of soil will begin to slide down. The wedge is defined by a plane that makes an angle β with the back of the wall. The area of the wedge, as shown in cross-section in Figure 2.10, multiplied by the unit weight of the soil gives the weight W of the wedge, so:

$$W = \gamma H^2 \tan (\beta) \qquad (2.3)$$

Do not confuse this W, the weight of the wedge, with W used to denote the weight of the wall itself. No equation should ever be taken from anywhere without checking its context and what the terms actually represent!

Because the full strength of the soil is being used, we can say that the shear force S acting on the back of the wedge is related to the normal force N according to: $S = N \tan \varphi'$. Now the forces acting on the wedge must be in equilibrium – if the wedge is not accelerating then the net force acting on it must be zero. A helpful way to show this is to use a force diagram, as shown in Figure 2.10. Each of the forces is drawn to scale, nose to tail, so that they form a closed polygon. The relationships can of course also be expressed using equations, and then the equations used to determine the value of P_a, the active pressure force provided by the retaining wall, which is needed to maintain the wedge in equilibrium. This is a matter of adjusting the angle of the wedge, β, until the maximum value of P_a is obtained. This gives the same expression for the active pressure force $P_a = K_a \gamma H^2/2$. This is known as a Coulomb wedge analysis, and may be extended to cover more complex situations, as given in Box 2.3.

2.4 LIMIT EQUILIBRIUM ANALYSIS

Once we know the pressure exerted by the retained earth on the back of the wall, we can assess its stability. In practice we will often be designing a new structure, so the problem becomes one of determining a suitable form and dimensions rather than just checking something that has already been defined. The essence of limit equilibrium analysis is to examine a potential failure mechanism. This is done by determining the forces that will make it more likely to occur if they are increased (the acting forces or actions), and those that will make it less likely to occur if they are increased (the resisting forces or reactions). Rather than attempting to determine the actual values of these forces, limiting values are used, and provided that equilibrium can still be achieved, that is, the resisting forces so calculated are greater than the acting forces, then the failure will not occur. The resisting forces are likely to be limited by the strength of materials, whereas the acting forces may be limited by reasonable choices about the loads to be designed for.

If we are considering the possibility of a retaining wall sliding forwards, then resisting force comes principally from a combination of the weight of

BOX 2.3   COULOMB'S EQUATION

Coulomb's equation (Figure 2.11) is an analytical solution to determine earth pressure for situations in which the back of the wall may not be vertical (at an angle α to the horizontal), the ground surface behind the wall may not be horizontal (sloping at angle i), and there may be friction (with an angle of friction δ) acting on the back of the retaining wall. The last condition is most important in drystone retaining walls, because the rough back face allows the full frictional strength of the soil to be developed, greatly improving the stability and efficiency of the structure.

$$K_a = \left\{ \frac{\sin(\alpha - \varphi)/\sin(\alpha)}{\sqrt{\sin(\alpha + \delta)} + \sqrt{\dfrac{\sin(\alpha + \delta)\sin(\varphi - i)}{\sin(\alpha - i)}}} \right\}^2 \qquad \text{(B2.3.1)}$$

Then

$$P_a = K_a \gamma H^2 / 2 \qquad \text{(B2.3.2)}$$

But note that this P_a is acting at an angle to the horizontal, so horizontal and vertical components need to be calculated for the analysis of the retaining wall.

$$P_{ah} = P_a \cos(\alpha + \delta - 90) \qquad \text{(B2.3.3)}$$

$$P_{av} = P_a \sin(\alpha + \delta - 90) \qquad \text{(B2.3.4)}$$

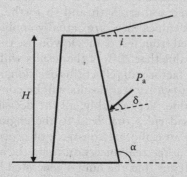

Figure 2.11 Parameters in Coulomb's equation for earth pressure.

the wall, and the frictional strength between the wall and the soil it is sitting on.

The acting force is a little more complex. What is usually done is to consider the acting force to be the earth pressure calculated previously, whereas in reality the acting force is just gravity pulling the retained soil downwards, and the frictional strength of this retained soil provides part of the resistance. The earth pressure is then calculated on the basis of using the full strength of the soil, and we concentrate on that part of the resistance that is proportional to the weight of the structure. The reasons for doing this are twofold; first, it takes very little movement for the full frictional strength of the backfill material to be developed; and second, when we look for a margin of safety, it is clearer if we are focussed directly on the effect of the decision we are making.

So what we aim for is that the resisting forces divided by the acting forces give a ratio that is greater than 1, which we call the factor of safety – 1.3 or 1.5 might typically be used. The simple clarity of the overall factor of safety raises questions about how the values to be used in the calculations have been determined. There is a sound argument for using values for resistance that we can be confident will be less than those that are actually available in a material that is variable. Correspondingly, we wish to use values for loads that we are confident will not be exceeded, whether due to the variable materials themselves or to varying traffic conditions on the ground the wall is retaining. If we are using such values, which can be described as conservative, then there is an argument for seeking a lower overall factor of safety, which perhaps takes into account the confidence in the accuracy of the analysis procedure, or the consequences should a failure occur. The problem with this is that it can be the top of a slippery slope – once you go down that route, you tend to go further and further, more and more quickly and out of control. This has happened in the engineering profession. The next small but reasonable step is to consider using partial factors of safety that are carefully chosen to reflect the uncertainty in each value that is used in the calculation – for the unit weight of the soil, the applied loading, for each component of each soil strength and so forth. Then there must be consideration of which values these are actually applied to, and how these are specified or obtained from test data. Do you use the average values of a range of results, or a value that 90% of the results will exceed? The appropriate value for partial factor will reflect this decision, and is often specified by a committee so that *typical* test results will be converted into a *typical* conservative design value. Very quickly, all the decision making becomes bound up in generalised rules that lead to the engineer becoming more and more focussed on an elaborate construct of procedures, and less and less likely to think about the real uncertainties in the actual problem being examined. The next step is that the hubris generated by the complexity of this procedure leads to a false confidence, and simplifications are made in the way the problem is thought about, which could have far greater consequences than any of these small decisions.

It is therefore much more secure to think carefully about the actions and reactions, and what might affect their values, taking care to make realistic decisions about how they might vary, before looking for an overall factor of safety. Many of the values used in the elaborate sets of partial factors are assembled by committees who are doing their best, but they are to an extent arbitrary, and so committees usually check that they produced similar answers to what had been obtained using overall factors – but inevitably only in a limited range of cases. The danger of this is that often the overall factors were developed over decades of use not to prevent a failure but to control deformations. Even if a failure does not occur, because of the ductile nature of most materials and designs, the deformation of a construction will increase significantly as the limit is reached, and the usual factors of safety were chosen to ensure that the most common materials and forms would still be in a region of behaviour where deformations were not excessive, and were in general elastic; that is, if you take a load off, the construction will go back to how it was before the load was applied.

If you were designing a retaining wall or a bridge foundation, and the 'live loading', that is, the load from vehicles that come and go, took the structure out of the elastic region, then each new application of load would push the structure a little further, from which it would not fully recover. So if a committee, or individual engineers, become too focussed on refining all those partial factors, they risk forgetting the need to control deformations.

The other danger is that the customary factors of safety provided insurance against something that really ought not to happen, but nevertheless occasionally did; if this is not explicit, a profession can forget the history that led to the practice, so that the historical failure is repeated.

The question of how to make good decisions about margins of safety using partial factors of safety is addressed in more detail in Chapter 4, but for now we will concentrate on considering what actually matters, while examining the limiting equilibrium of a simple gravity retaining wall.

2.4.1 Sliding

The configuration we will consider is shown in Figure 2.5b, together with the forces acting. If we describe the wall height as H and its width as B, and assume the cross-section to be a simple rectangle, then we can describe the forces per unit length of wall.

The weight of the wall:

$$W = \gamma_w HB \tag{2.4}$$

The reaction force Q on the base of the wall is simply equal to the weight of the wall, as we are taking the friction on the back of the wall to be zero.

The horizontal component of the total force from earth pressure on the back of the wall, which is the sole acting force, or action, is

$$P_{ah} = K_a\gamma_s H^2/2 \tag{2.5}$$

in which, for zero friction on the back of the wall and a vertical wall back

$$K_a = \frac{1-\sin(\varphi')}{1+\sin(\varphi')} \tag{2.6}$$

And as the friction on the back of the wall is zero, the vertical component of earth pressure, $P_{av} = 0$.

The frictional strength on the base, which provides the resisting force or reaction is

$$T = Q \, \alpha \tan(\varphi') = \gamma_w HB \tan(\varphi') \tag{2.7}$$

where α is a coefficient of interaction, which accounts for the fact that the base of the wall might not be as rough as the soil itself, and so the friction between wall and soil is less than the frictional strength within the soil. The value will therefore typically not be less than 0.5 and cannot be more than 1.0. In practice, efforts would be made to produce a rough base and a higher value would be appropriate.

Then the factor of safety against sliding is just T/P_{ah}:

$$F_{sliding} = \frac{\gamma_w HB \tan(\varphi')}{K_a\gamma_s H^2/2} \tag{2.8}$$

As an example, consider a wall 5 m high and 2 m wide, made of concrete with a unit weight γ_w of 24 kN/m³. The retained fill has a unit weight $\gamma_s = 20$ kN/m³ and an angle of friction $\varphi'_s = 35°$.

The wall sits on soil with a friction angle of $\sigma'_f = 30°$, and the relatively smooth concrete base has an interaction factor $\alpha = 0.5$.

Then using three significant digits:

$$K_a = \frac{1-\sin(\varphi')}{1+\sin(\varphi')} = \frac{1-\sin(35)}{1+\sin(35)} = \frac{1-0.574}{1+0.574} = 0.271 \tag{2.9}$$

and so the acting force is

$$P_{ah} = 0.271 \times 20 \times 5^2/2 = 67.8 \text{ kN/m} \tag{2.10}$$

The weight of the wall is

$$W = 24 \times 5 \times 2 = 240 \text{ kN/m} \qquad (2.11)$$

and gives a resisting force:

$$T = 240 \times 0.5 \times \tan(30) = 69.3 \text{ kN/m} \qquad (2.12)$$

It can be seen right away that the resisting force is only just greater than the acting force, giving a factor of safety against sliding only just greater than 1:

$$F_{\text{sliding}} = 69.3/67.8 = 1.02 \qquad (2.13)$$

So in theory this wall would be acceptable, but in practice we would want a larger margin of safety to account for the uncertainties in the data, and to give those who might be affected by the wall some real security! As the limit is reached, movements would become significant, but the consequence of such movement would be that the fill behind the wall would drop down, so reducing the pressures and allowing the wall to stabilise. So a sliding failure would be unlikely to be catastrophic, unless the effects of settlement of the fill behind the wall were serious, and a normal value for factor of safety would be 1.5.

This leads to the idea that instead of simply analysing a guessed proposal, the engineer should use the mathematics to predict a wall geometry that would achieve a required factor of safety with the materials available. In this case we would wish to achieve $F_{\text{sliding}} = 1.5$, and hence $T = 1.5 \times P_{\text{ah}} = 101.6$ kN/m. This would require either an improvement in α or an increase in W, or both. So if we assume that the base of the wall can be made to achieve $\alpha = 0.75$, we then need $W \times 0.75 \times \tan(30) = T = 101.6$ kN/m. Hence, W must be 234.6 kN/m, which is less than the weight we have already, because we have improved the friction on the base by 50%, and that gives enough margin of safety. In fact, ensuring reasonable frictional resistance on the base is not difficult.

The next stage in making this assessment more realistic is to consider the effect of friction on the back of the structure, which produces a vertical component to the earth pressure force, as shown in Figure 2.5. A typical decision would be to use $\delta = 2\ \varphi'/3$, though some would prefer to define this in terms of the achieved coefficient of friction, so that $\tan(\delta) = 2\tan(\varphi')/3$, the difference between the two approaches being insignificant in comparison with the uncertainty in the value. This leads to the more complicated determination of K_{a} as shown in Box 2.3, giving a value of 0.232 for $\delta = 23.3°$ and $\varphi' = 35°$. This results in $P_{\text{ah}} = 53.4$ kN/m and $P_{\text{av}} = 23.0$ kN/m. This means that we need $T = 1.5 \times 53.4 = 80.1$ kN/m. Therefore, Q must be 185 kN/m, but Q is now made up of the combination of W and P_{av}, so W need only be 162 kN/m, requiring $B = 162/(5 \times 24) = 1.349$ m. Compared with the original 2 m, it can be seen that this is 30% more efficient in its use of material (and hence more sustainable and better engineered).

Considering the vertical and horizontal components of earth pressure separately makes the calculations easier, but can be misleading when assessing a factor of safety, especially if partial factors are used. This is because in assessing safety, whether using partial factors or overall factors, we should be looking at things that really could vary independently. Thus, in looking at sliding, the retained fill could be a little more dense than assumed at the same time as the foundation soil being a little weaker than assumed, but the horizontal component of the earth pressure and the vertical component are just two parts of the same force, and are not independent of each other. The relationship between them, however, is governed by the wall friction δ, so if we want to consider an unfavourable value of P_{av} (low) at the same time as an unfavourable value of P_{ah} (high), we should really be using a minimum value of delta. However, although this might give an increased margin of safety, it would be poor engineering unless the δ we use is really a possibility.

2.4.2 Bearing and overturning check

Assessing the bearing capacity of the ground the wall is sitting on requires specialist investigation and analysis, but is covered by many standard textbooks on soil mechanics and geotechnical engineering. We are going to concentrate here on the distribution of pressure exerted by the retaining structure on the ground, which is closely related to its resistance to overturning. First, a simple check of the overturning mechanism may be made.

Overturning is a rotation (Figure 2.6b), and so we analyse in terms of moments. We are considering whether or not the structure will rotate about its toe, and therefore we take moments about the toe. This requires us to consider each force acting on the structure, including its self-weight, and determine its point of action and direction, and so determine its 'lever arm' about the axis of rotation. These concepts are shown in Figure 2.12a. The moment is then the force multiplied by the lever arm. The limit equilibrium check is then a comparison of the moments which cause rotation with the moments that can resist rotation, and we would normally seek an overall factor of safety for this. As the limit is reached, the stresses placed on the wall material at the toe, and on the foundation beneath it, would reach a level that would cause them to fail, but this would usually be preceded by sufficient movement to cause alarm and prompt investigation. If significant settlement took place under the toe of the structure, then it would end up leaning forwards, so the resisting moment from its own weight would be reduced and this could lead to an accelerating failure, or catastrophic collapse. The realisation that something bad would probably happen *before*

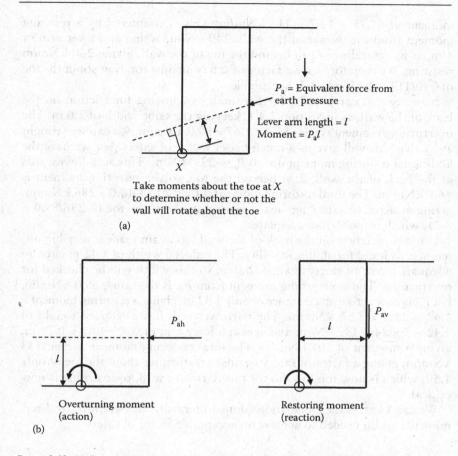

P_a = Equivalent force from earth pressure

Lever arm length = l
Moment = $P_a l$

X

Take moments about the toe at X
to determine whether or not the
wall will rotate about the toe

(a)

P_{ah}

l

Overturning moment
(action)

P_{av}

l

Restoring moment
(reaction)

(b)

Figure 2.12 (a) Force, moment and lever arm. (b) Components of active pressure force.

the limit is reached means that engineers would usually look for a higher factor of safety than for sliding; most commonly 2.0 would be sought.

Though Figure 2.12a shows that the earth pressure force can be assessed as it is, producing a moment that causes overturning to take place, it is common to separate out the horizontal and vertical components of this force, as was done when considering sliding, and as shown in Figures 2.4 and 2.12b. In this case, the horizontal component produces the main acting moment, while the vertical component produces a resisting moment, which is additional to the main moment coming from the weight of the retaining wall. Because the effect of the vertical component is significant, it is worth looking at the case for a smooth wall back ($\delta = 0$) that we started with when examining sliding resistance.

So considering the moments that make the wall tip forward, we have just $P_{ah} = 67.8$ kN/m, which acts at height $H/3 = 1.67$ m, giving an overturning moment of $67.75 \times 1.67 = 113$ kNm/m. This is countered by a resisting moment from the weight of the wall, 240 kN/m, acting at a lever arm of 1 m, as its centreline is 1 m behind the toe of the wall, giving 240 kNm/m resisting. We therefore have a factor of safety against rotation about the toe of $240/113 = 2.13$, which is acceptable.

Now we can carry out a further analysis allowing for friction on the back of the wall, as in Section 2.4.1, keeping the same width of 2.0 m. The overturning moment is now $53.4 \times 1.67 = 89.0$ kNm/m. We can see straight away that this will give us a much better factor of safety, but we have the additional restoring moment due to $P_{av} = 23.0$ kN/m. This acts downwards at the back of the wall, 2 m behind the toe, so the restoring moment is 46.0 kNm/m. The total restoring moment is then $240 + 46.0 = 286$ kNm/m, giving a factor of safety against overturning about the toe of $286/89.0 = 3.21$, which is more than adequate.

Using the friction on the back of the wall has again made a very big difference, as it did for sliding stability. The reduced width of 1.42 m gave an adequate factor of safety against sliding, so this width will be checked for overturning. The overturning moment from P_{ah} is the same, 89.0 kNm/m, but P_{av} is now acting at a distance of only 1.42 m, giving a restoring moment = $1.42 \times 23.0 = 32.7$ kNm/m. The narrower wall has a reduced weight of $1.42 \times 5 \times 24 = 142$ kN/m, and acts at a lever arm of $0.5 \times 1.42 = 0.71$ m, giving a moment of 101 kNm/m. The total restoring moment is then 133 kNm/m, giving a factor of safety against overturning about the toe of only 1.50, which is now too low. So for this narrower wall, overturning is now critical.

We can express the calculations done in terms of the width B, to determine the width needed to achieve an acceptable factor of safety:

$$F_{overturning} = 2.0 = \frac{P_{av}B + 0.5BBH\gamma_w}{P_{ah}H/3} \tag{2.14}$$

This may be rearranged to give

$$2P_{ah}H/3 = P_{av}B + 0.5B^2 H\gamma_w \tag{2.15}$$

And then to

$$0.5B^2 H\gamma_w + P_{av}B - 2P_{ah}H/3 = 0 \tag{2.16}$$

This is a quadratic equation which can be solved using the usual formula. Entering the numbers:

$$0.5B^2 \times 5 \times 24 + 23.0B - 178 = 0 \tag{2.17}$$

which is

$$60B^2 + 23.0B - 178 = 0 \qquad (2.18)$$

The solutions are then

$$\frac{-23.0 \pm \sqrt{(23.0^2 + 4 \times 60 \times 178)}}{2 \times 60} \qquad (2.19)$$

The positive solution is then 1.54 m. This width of structure would be satisfactory in terms of both sliding and overturning, but we would also like to know if the pressure distribution on the base is reasonable.

This is examined by considering the position of the resultant force on the base – the single force which is equivalent to the pressure acting across the width of the base, and is in equilibrium with the vertical forces acting downwards on the foundation soil. To maintain moment equilibrium as well as force equilibrium, this force needs to be acting in front of the centreline of the base by a lever arm known as the eccentricity e (Figure 2.5b). Then the net moment acting on the base M is resisted by the vertical force Q multiplied by the eccentricity e. Note that it is most convenient to consider rotation about the centreline of the base when considering bearing pressure – this is a different assessment from checking overturning, though there is a relationship between the two assessments as will be explained in the following paragraphs.

If there is no net moment about the centre of the base then the pressure distribution will be uniform. It is possible to design an earth retaining structure that will work like this, by having most of the weight towards the back – so it leans back against the soil. However, the simple rectangular section we are considering, with the earth pressure acting on it, will apply a moment to the base; hence the pressure beneath the toe will be greater than the pressure beneath the heel (Figure 2.13a). If the earth pressure is higher, then this difference in pressure will increase (Figure 2.13b) until the pressure beneath the heel drops to zero (Figure 2.13c). As this increase has been happening, the position of the resultant force on the base has moved forwards from the centreline, and is now one third of the way to the front of the base, so that the eccentricity = $B/6$, and it is said that the resultant is just within the middle third.

The next step is important to understand. Because it is commonly said that soil cannot carry tension, some might say that if the moment increases, then the resultant will no longer be within the middle third, and the soil will fail because it cannot carry tension. This 'failure', being a failure, is regarded as unacceptable, and so the resultant must be kept within the middle third – they say. However, all that actually happens is that pressure is no longer applied over a strip at the back of the base (Figure 2.13d),

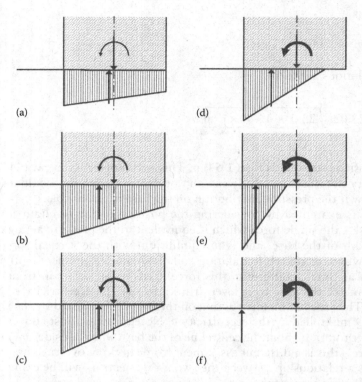

Figure 2.13 Bearing pressure – the 'middle third' rule. (a) If the earth pressure on the back
of the wall is light compared with the weight of the wall, the moment about
the centreline of the base will be relatively small, and the pressure distribution
is only a little greater beneath the toe. (b) As the earth pressure increases, the
pressure beneath the toe becomes significantly larger, and the toe will sink a
little deeper than the heel. (c) If the earth pressure is high enough, the pressure beneath the heel drops to zero and the pressure distribution becomes
triangular. At this point, the eccentricity of the resultant is one sixth of the
base width – the resultant is only just within the 'middle third' of the base.
(d) Once the earth pressure passes this value, some engineering scientists
would say that the wall will fail because the soil cannot carry tension. All that
really happens is that the back of the triangle will shift forward, so that pressure is not applied over the full width of the base. (e) However, as the moment
gets larger, the pressure on the base acts over a strip which gets narrower,
so the pressure becomes very large, and the foundation will eventually fail.
(f) When the moment is large enough, the whole vertical load is supported by
a line at the toe, and the structure is on the point of overturning.

and the triangular distribution of pressure will be applied over a steadily
reducing width as the moment gets larger and larger. At the same time,
the maximum pressure will be increasing, and when this becomes a problem depends on the strength and stiffness of the foundation (Figure 2.13e
and f). There will come a point with any foundation material when it will

fail, because when the resultant reaches the toe the pressure will be infinite, as the entire load is acting on a line of zero width. It is possible that the material of the wall itself might crush first if the foundation is very strong.

The triangular pressure distribution could also lead to a compression of the foundation that leads to the wall tilting forwards significantly – this would at least mean that people would be worried about the structure, and may have it investigated before a failure occurs, but it is not a configuration to be deliberately designed. So the main thing to note is that when the resultant passes outside the middle third, nothing happens. Depending on the strength of the foundation, if the resultant passes a long way outside the middle third, then there may eventually be a bearing pressure failure.

If the resultant is so far outside the middle third that it reaches the front of the base (the toe), and the foundation and the lowest part of the wall are infinitely strong, then the wall will be on the point of rotating forwards about the toe – an overturning failure. This situation will be prevented by ensuring an adequate factor of safety in the overturning check, and it will also be prevented if the resultant is kept within the middle third. It is possible that the strength of the foundation is inadequate even if the resultant is within the middle third, and a bearing capacity check needs to be done anyway – but keeping the resultant in the middle third will help. Because nothing actually happens when the resultant reaches the middle third (it is not a 'limit state'), this check should be done with unfactored design loads and soil strengths, to assess what could actually occur, and without incorporating any 'margin of safety'. If the middle third criterion is satisfied, there is automatically a margin of safety against bad things happening; it is just a criterion that results in a reasonable design.

2.4.3 Effects of varying the geometry

The retaining structure described earlier works through its weight, generating frictional resistance to prevent forward sliding, and resisting the overturning forces of the earth pressure acting on its back. Variations in the geometry from the simple rectangular section considered earlier can help these functions, but their impact on the difficulty of construction and the aesthetics may also be significant.

The simplest variation is to incline the front face backwards, as shown in Figure 2.14a. This has the benefit of moving the centre of gravity of the wall backwards, so that its weight acts behind the centreline, providing a restoring moment to help resist the overturning moment from the earth pressure, and so bring the resultant force Q closer to the centreline of the base. The result will be a more uniform distribution of pressure on the foundation. This configuration also has aesthetic benefits. It is instinctive to expect something which is holding back the earth to be leaning back against it, so the battered front face simply looks right. When the effects of settlement are taken into account this

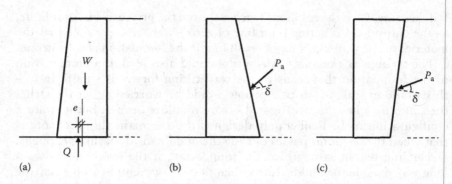

Figure 2.14 Variations in wall geometry. (a) Front face inclined backward. (b) Back face inclined forward. (c) Back face inclined backward.

becomes more important. The pressure on the foundation will normally be greater towards the toe, so that the wall will lean forwards a little as the soil beneath it compresses. The movement will normally be slight, but if the face is battered back when constructed, then it is much less likely that the wall face will end up leaning over forwards, which to most people would be alarming, even if the structure was actually safe and stable because of its thickness. Compared with a rectangular cross-section wall, a wall with a battered face using the same volume of material will be using that material more efficiently to resist overturning, but the sliding resistance will be unaltered.

If the back face is inclined forwards, as shown in Figure 2.14b, then the earth pressure changes. The benefits of friction on the back of the wall, in resisting overturning and generating more sliding resistance, are augmented by the normal force having a vertical component and by the shear force acting more downwards than before. However, the overall force P_a increases, and the effects of this are not offset by the increase in P_{av}. As an example, if the back of the wall analysed previously, with $\delta = 23.3°$ and $\varphi' = 35°$, is inclined by just 10°, then K_{ah} increases from 0.214 to 0.246, and K_{av} from 0.092 to 0.162. This adds 8.09 kN/m to P_{ah}, but 17.4 kN/m to P_{av}. The increase in P_{av} adds 7.5 kN/m to the sliding resistance, nearly offsetting the increase in P_{ah}, but the factor of safety will reduce. A fair comparison would be for the same volume of wall material, so the weight of the wall is the same, but even so, the 10° batter has removed about 53 kN/m from the back of the wall that has not been taken from the front, which will reduce the restoring moment, and the increase in P_{av} will not make up for it.

If the back face is instead inclined backwards, then the weight of the wall, and the consequent sliding and overturning resistance, are affected in the same way as for the case above, but in terms of moment about the centreline of the base, things look better because the centre of mass of the wall is behind the centreline, rather than in front of it. In this case K_{ah} decreases from 0.214 to 0.174, and K_{av} from 0.092 to 0.041. This removes

9.8 kN/m from P_{ah}, and 12.7 kN/m from P_{av}. The reduction in P_{av} only takes 5.5 kN/m from the sliding resistance, so if the overall weight of the wall is kept the same, the factor of safety against sliding will improve slightly. The effect on restoring moment of the reduction in P_{av} will be more than offset by the more favourable geometry of the wall.

This assessment leads to the conclusion that more efficient use will be made of the wall materials if both front face and back face are inclined back towards the fill – a conclusion that might be reached intuitively. Structures of this type exist and are called *perré* in French; if made from drystone, they are referred to as drystone revetments, and they are used to stabilise and protect earth and rock dams for example. However, in practice the backwards leaning wall is more difficult to construct, as either the soil must be placed behind it as it is built, or it must be built overhanging at the back, and the soil compacted beneath that overhang. It is therefore most common to construct the back face vertical.

2.4.4 The effect of loading on the ground surface

A retaining wall is usually made to provide a level space in front of the wall or on the fill behind it. In any event, there will be some construction equipment on the fill behind the wall at some stage, and commonly there is a road. The loading from the road must be accounted for in the design. This is usually done by allowing for a uniform surcharge across the ground surface behind the wall, even though actual loads are unlikely to be uniform, coming typically from wheels. The effect of a uniform load is to increase the vertical stress, which as seen earlier will result in a proportional increase in horizontal pressure on the back of the wall. If the surcharge is expressed as Q (normally in kN/m^2), then the resulting horizontal pressure will be $K_{ah}Q$, and if there is friction on the back of the wall, there will also be a shear stress of $K_{av}Q$. These stresses will act over the full height of the wall, resulting in additional forces $P_{ahq} = HK_{ah}Q$ and $P_{avq} = HK_{av}Q$. These forces are easily incorporated into the calculations. A typical value for Q might be 10 kN/m^2, which is only equivalent to slightly more than 0.5 m of soil. However, the effect of the actual concentrated wheel loading close to the top of a wall can be significant if the structure is made of drystone, as will be seen in Chapter 3.

2.5 YIELD ANALYSIS

For the simplest earth pressure calculations, considering the stresses behind the wall quickly leads to the same answers as considering potential failure mechanisms, as noted in Section 2.3. This coincidence can be put into a broader theoretical context by examining limit analysis, a theoretical tool originally developed for studying the deformation of metals and also described as 'plasticity analysis'. At the time this was done, clay soils were treated as purely cohesive, that is, they have a shear strength which is independent of the normal stress. They appear to behave like this for short-term or rapid loading because of the effects of water pressure, even though their true behaviour is frictional. The metals that were being analysed also showed a simple yield strength, and so the theoretical work done on metals could be transferred quite easily to 'cohesive' soils. This yield strength is a measure of the stress, and if the material is 'perfectly plastic' then it will continue to yield, or deform, as long as that stress is maintained. In very particular circumstances, some real soil can actually behave like that. Applying the theories to real frictional soil was not straightforward and required assumptions to be made which are not generally true. However, using these theories can give useful insights into real problems, and provided that the engineer gives proper consideration to the differences between the theory and the reality (a normal requirement for an engineer), they can even be put to use in practice.

When we explored the relationship between horizontal and vertical stress at yield, the picture we developed was what is described as a compatible stress field. That is, the described stresses could exist, and be in equilibrium. This is described as a lower bound approach, because in general for a perfectly plastic material the actual load required to cause collapse cannot be less than the calculated value. Though real soil is not a 'perfectly plastic material', if deformations are kept very small, as may be desired in a working structure, then the error in making this assumption may also be small. There is also an assumption about the relationship between patterns of stress and patterns of deformation called the 'associated flow rule'. This means that the direction of plastic strain is 'normal to the yield surface', which in turn means that as frictional soil shears, it expands at the friction angle. A material that is purely cohesive, that is, has a shear strength that is not affected by the normal stress, must not expand or contract as it shears. The proofs of the upper and lower bound theorems require an associated flow rule.

The yield design theory states that if $G(\underline{x})$ is a convex domain given in the six–dimension vector space of the stress tensor $\underline{\sigma}(\underline{x})$, which defines the strength resistance, then a condition for the system $\overline{\Omega}$ to remain stable is

$$\mathrm{div}\,\underline{\sigma} + \rho(\underline{F} - \underline{\gamma}) = 0 \ \ \text{equilibrium}$$

$$\underline{\sigma} \cdot \underline{n} = \underline{T}^{\mathrm{d}} \ \ \text{over}\ \partial\Omega\ \text{boundary conditions}$$

$$\underline{\underline{\sigma}} \in G(\underline{x}) \ \ \text{strength criterion}$$

where ρ is the material density, $\rho\underline{F}$ the body forces applied over Ω, $\underline{\gamma}$ the acceleration field developed in Ω and $\underline{T}^{\mathrm{d}}$ the static boundary conditions applied over $\partial\Omega$ (Figure 2.15).

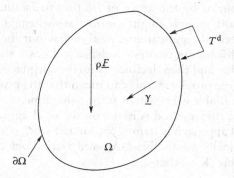

Figure 2.15 Mechanical system.

By writing the preceding equations, an estimation of the limit loading could be computed. Two estimation approaches may be implemented.

Lower bound analysis corresponds to the stress-based approach, whereas the velocity field-based approach corresponds to upper bound analysis. The upper bound and lower bound theorems, at face value, give some results that are certain. However, this is only from a theoretical point of view, and the good geotechnical engineer is always very aware that not only are the differences between the theoretical and the real behaviour important, but also the variability of the material properties can introduce a much more significant uncertainty in the results of any analysis.

The lower bound theorem proves that for very specific assumptions about material behaviour, if a 'solution' is found that shows a stress field in equilibrium with applied loads, then there will definitely not be a collapse under a lesser load. This is described in French more graphically: *statique par l'intérieur* (Salençon 1983, 2013). It is looking at a static situation, from within the body of the material. This approach is not used by the authors for the design of the structures in this book.

The upper bound theorem proves, for the same very specific assumptions about material behaviour, that if a 'solution' postulates a mechanism in which the rate of work done by the applied loads equals or exceeds the work done within the deforming material, then that mechanism may take place, but there may be a mechanism that can occur under a lesser load. The calculated load is therefore an upper bound to the collapse load, and it may be necessary to try to find other mechanisms that move more easily. This is described much more usefully in French as *cinématique par l'extérieur* (Salençon 1983, 2013). It is looking at a kinematic situation, from the point of view of what is happening on the boundaries of the material. It is important that the mechanism has to be kinematically admissible – thus relative movement between two defined bodies of soil can only be in shear, that is, along the boundary. However, when soil shears it usually expands as the soil particles move up and over each other, and this expansion is presumed to be at a rate defined by the angle of friction to facilitate the plasticity analysis. A loose soil, though, might contract as shearing takes place. These differences between theoretical and real behaviour become important only if a lot of deformation occurs – a dense soil has a shear strength that reaches a peak value and then declines as further displacement takes place. Particularly for large problems, this can mean that the peak strength of the soil may not be available everywhere at the same time.

Mathematically, this method is based on the dual approach of the equilibrium. This dual approach is named the virtual work equation. It says for any stress $\underline{\underline{\sigma}}$ statically admissible (SA), and virtual velocity field $\underline{U}$ kinematically admissible (KA), that

$$\int_{\Omega} \underline{\underline{\sigma}} : \underline{\underline{d}}\, dV + \int_{\Sigma_u} \left(\underline{\underline{\sigma}}[U]\right)\underline{n}\, dS = P_e\left(\underline{\underline{\sigma}}, \underline{U}\right)$$

where P_e is the virtual rate of work by all the external forces in equilibrium with $\underline{\underline{\sigma}}$ in the virtual velocity field $\underline{U}$ and Σ_u is any optional surface where the velocity field is discontinuous with discontinuity of $\underline{U}$ noted $[U]$. The yield design theory, which could be written, if K is the set of admissible loading, as

$$\underline{Q} \in K \Leftrightarrow \exists \underline{\underline{\sigma}} \left\{ \begin{array}{c} \underline{\underline{\sigma}}\, SA \text{ with } \underline{Q} \\ \underline{\underline{\sigma}}(\underline{x}) \in G(\underline{x})\ \forall \underline{x} \in \Omega \end{array} \right\}$$

leads to

$$\underline{Q} \in K \Leftrightarrow \exists \underline{\underline{\sigma}} \left\{ \begin{array}{c} P_e\left(\underline{\underline{\sigma}}, \underline{U}\right) = \int_{Omega} \underline{\underline{\sigma}} : \underline{\underline{d}}\, dV + \int_{\Sigma_u} \left(\underline{\underline{\sigma}}[U]\right) \cdot \underline{n}\, dS \\ \underline{\underline{\sigma}}(\underline{x}) \in G(\underline{x})\ \forall \underline{x} \in Omega,\ \forall \underline{U}\, KA \end{array} \right\}$$

If one defines the following π functions as the upper bounds of the quantities integrated in the right-hand side of the preceding equality as

$$\pi(\underline{\underline{d}}) = \begin{array}{c} Sup \\ \underline{\underline{\sigma}} \in G(\underline{x}) \end{array} \left(\underline{\underline{\sigma}} : \underline{\underline{d}}\right)$$

$$\pi(\underline{n}, [U]) = \begin{array}{c} Sup \\ \underline{\underline{\sigma}} \in G(\underline{x}) \end{array} \left(\underline{\underline{\sigma}}[U]\right) \cdot \underline{n}$$

where $\underline{n}$ defines the surface of discontinuity of $\underline{U}$ $\underline{n}$ being the normal vector of the surface. Then kinematic approach of the yield design theory is defined by

$$\underline{Q} \in K \Rightarrow \left\{ \begin{array}{c} \underline{Q}\dot{\underline{q}}(\underline{U}) \leqslant \int_{\Omega} \pi(\underline{d})\,dV + \int_{\Sigma_u} \pi\left(\underline{n},[\![\underline{U}]\!]\right)dS \\ \forall \underline{U}\ KA \end{array} \right\}$$

where it may proved, with the generalised virtual velocity of the system, $\dot{\underline{q}}(\underline{U})$, that

$$P_e = \underline{Q}\dot{\underline{q}}(\underline{U})$$

With the study of a few kinematically admissible velocity fields (mechanisms), an upper bound value of the extreme load that the structure could support can be obtained.

Care is needed in thinking about loads in a frictional material – the same load may be acting for one mechanism and resisting for another. In any case, an externally applied load may be resisting the effects of self-weight of the soil, so great clarity is needed in thinking about upper bound and lower bound – no mathematical expression should be used unless it is fully understood. Engineering scientists sometimes make much of the fact that these theorems can be 'proved' mathematically, but a verbal explanation is far more useful and convincing, especially so for soil that does not satisfy the underlying assumptions anyway.

For the lower bound: If you were to set up the soil with the pattern of stress that you postulate, and yield begins, then the effect will be to transfer load within the structure/body of soil from highly stressed regions to less highly stressed regions, thus lowering the peak stress and stopping yield so that the load can be increased further. This is not the case, of course, if you have found the critical stress state, where there is no scope for redistributing stress, and your lower bound coincides with the upper bound.

For the upper bound: If you have shown a mechanism for which more work is done by the forwards movement of acting forces than is expended on pushing resisting forces back, then there will be a surfeit of energy that will result in acceleration. So if it can move, it must move, and it will move faster until the geometry or loading have changed sufficiently for the balance of forces to change to inhibit movement, the kinetic energy is reabsorbed and balance restored. But there might well be mechanisms that can move more easily. So every analysis you do represents an upper limit on the load that can be carried, and you must continue doing analyses until you are confident that you have found the lowest upper limit, or are close enough to it (unless of course you just wanted to find any load that would guarantee movement). If you have already tried hard to find the highest lower bound, and this coincides with or is reasonably close to your lowest upper bound, then you will conclude that you have a good enough assessment of what might really happen.

In both cases, you can see that when expressed like this the essence of these ideas must be correct even if the soil is not perfectly plastic. The difficulty lies in the detail. These theorems both require that either an entire body of soil (lower bound) or failure surface (upper bound) can be at yield stress at the same time, implying perfect plasticity. This is the essence of the idea of the 'critical state' (described in most soil mechanics textbooks), but soil is not normally at the critical state, and so does not behave in this way.

However, soil is usually strain hardening (i.e., it gets stronger as it shears) before failure (if it is overconsolidated), and then strain softening (i.e., it gets weaker as deformation continues). It therefore matters a lot how the situation you are analysing developed – and when you consider that soil consolidates under increasing normal stress, during which process its strength and deformability change, you will realise that this simplistic way of looking at things must be getting more wrong than right. In particular, in a large mechanism in compressible soil, one part of a mechanism may have passed peak before another part has reached peak – this is called progressive failure. And thanks to the strain softening, as soon as one part of the soil reaches yield, it will weaken and throw load onto adjacent soil so that a shear surface develops. This means that shearing becomes concentrated within a very thin zone, and the 'associated flow rule', which relates patterns of strain to patterns of stress, is no longer valid.

The associated flow rule might be credible at the onset of yield – for example, it implies that soil in a slope at the angle of friction is stable because a failing element of soil would in effect have to move horizontally, not down the slope. This implies that there is no intrinsic friction between soil particles, only a geometrical effect due to the dilation as shearing takes place. A slight steepening of a slope beyond the angle of friction would result in the soil moving downwards. Such dilation can in reality only be very transient, so an attempt to model the kinematics using this assumption is bound to be totally unrealistic. The effect, however, is to impose a kinematic behaviour on the soil that makes it look like the simple elastic-perfectly plastic metals for which plasticity theory was originally devised, and from which theoreticians wished to borrow results. In fact it is possible to investigate the kinematics using a dilation angle based on observation, just not as easy. The simple results are for dilation = friction, and the theoretical neatness appeals to the academically minded, whereas the practically minded do not want to ignore real soil behaviour.

One of the consequences of blind adherence to plasticity theory is that very academic engineers prove that the only possible curved failure surface for a frictional soil is a logarithmic spiral, and you will see this used in some theoretical work. A logarithmic spiral can in fact be representative of the behaviour of soil at very small deformations, but it is important to realise that it is an artefact of unrealistic assumptions, not the 'true' behaviour, when you read textbooks or academic papers.

Limit equilibrium analysis looks a great deal like upper bound analysis, but it is possible to get at an answer without testing the feasibility (or kinematic admissibility) of the mechanism. If you carry out a limit equilibrium analysis without taking into account kinematic constraints, then there may be more work to be done than you have accounted for if any movement is actually to take place on your mechanism, and so a larger load will be needed than you have calculated. So your result is not really an upper bound, because you have not taken into account the kinematics. This would usually result in a conservative design – the construction is stronger than you think. On the other hand, if you have a soil that collapses rather than dilates, the kinematics actually make failure easier, and the design will be weaker than you think. The most important thing is to know which is happening. If you can take account of the kinematic admissibility to some extent in your limit equilibrium analysis, and make the mechanisms you analyse more realistic, then you will be getting closer to an upper bound analysis and so a design that is not unduly conservative.

For the analysis of retaining walls, especially of gravity walls, the fill behind the wall will usually be dense enough that it dilates on shearing, so that the limit equilibrium analysis will be conservative. However, thinking about stresses and mechanisms in a careful way, as required by yield analysis, can give useful insights into the modes of behaviour. In either case, the analysis is unlikely to introduce a possible mode of failure if the engineer has not already thought of it, so imagination and visualisation are very important.

2.6 SUMMARY

Gravity retaining walls have been used for thousands of years, and have mostly been made of drystone construction. Many walls have been built to proportions that seemed right to the builder, and were proven to work by experience. Methods of analysis for gravity walls are now well established, and design is now based on those methods of analysis, to ensure satisfactory margins of safety. However, increasing conservatism in engineering practice, culminating in international, committee-produced codes of practice, can make it difficult for the practicing engineer to realise just how efficient a gravity wall can be – especially if it is drystone. The unique features of drystone retaining wall behaviour are explored in Chapter 3, with a view to supporting efficient, sustainable design, analysis and construction.

Chapter 3

Behaviour of drystone retaining structures

The discussion of simple earth retaining structures in Chapter 2 concerned the relationship between the weight of the structure and the surrounding soil. This chapter considers the effect that particular characteristics of drystone masonry have on that behaviour. A drystone wall is in many ways the ideal gravity retaining wall, having good permeability to prevent the build-up of pore water pressures, strength where it is needed and excellent ductility.

3.1 TRANSMISSION OF FORCES WITHIN A DRYSTONE WALL

The stones within a drystone wall may move relative to each other, as they are just resting on top of each other. When a stone is placed during the building of a wall, its weight will initially rest on only three points. The builder will try to adjust the position of the stone, or use a wedge, so that is supported on a fourth point and does not rock. A securely placed stone will remain in this position, but it may be moved slightly as adjacent stones are placed, or stones laid on top of it, so that it is transmitting load to the stone or stones below through only one or two points. In this case it is prevented from rotating by the newly placed stones. However, most stones will be supported at their corners – three for a triangular piece and four for a rectangular piece – and will be stable. This is easily seen and understood if the stones are generally flat, but rounded stones can be much more difficult to place and depend on smaller stones placed around them for their stability. In any case, forces are transmitted through the wall only via the points where stones are in contact with each other.

The points of contact between stones can transmit direct forces and shearing forces. For most stones used in drystone wall construction, the angle of friction at the points of contact is high, as is the strength of the rock to carry compression forces. It is therefore likely that the wall will slide on its base rather than on a plane that passes in between its stones. If the foundation is very strong, then sliding between the stones may occur.

3.2 THE EFFECT OF LOADING AT THE GROUND SURFACE

Loads applied on the backfill may have more significance for a drystone wall than for a concrete gravity wall, because elements of the wall can move independently. Earth pressure from a surcharge load is constant with depth, and the equivalent force acts at half the height of the wall; as with the earth pressure due to the self-weight of the soil, the horizontal component will produce an overturning moment, and the vertical component arising from friction on the back of the wall will produce a resisting moment. Unless the wall is small, the moments are not large in comparison with those due to the weight of the soil, but they nevertheless result in a significant difference in what the wall has to do. The real loading is unlikely to be a uniform load across the whole area – it is more likely to be a series of concentrated wheel loads. For a mass-concrete or cemented masonry wall, the difference may not be significant, as it is the averaged-out loading that matters, but for drystone walls the concentrated loads could cause localised damage. It becomes important to know how close a wheel might come to the top of the wall, as if it is not too close then the load spreads out through the backfill with depth so the stresses are not as high, and it affects the wall deeper down where it can carry additional stresses more easily. A drystone wall is capable of redistributing loads within it, giving it a capacity to deal with concentrated wheel loads, but this ability is highly dependent on the quality of the construction, as explained in Sections 3.5 through 3.7.

3.3 THE EFFECT OF THE ROUGH BACK FACE

The single most important difference between a drystone retaining wall and a simple mass concrete or masonry gravity wall is the roughness of the back face, which enables the full friction of the backfill soil to be developed on the back of the wall. As explained in Chapter 2, this force greatly assists the stability of the wall and allows drystone structures to be quite slender. To appreciate just how large an effect this can have, an example is shown in Figure 3.1a. This assumes that the backfill is a good quality granular material with an angle of friction of 38°, and that this full friction angle acts on the back of the wall. As can be seen, a base width of 2.1 m is sufficient to keep the force from earth pressure due to the weight of the backfill acting within the width of the base; in other words, in the configuration shown, the earth pressure from the soil produces zero overturning moment about the toe. This allows the wall to be very much more slender than if the backface was not rough, and this wall would have acceptable factors of safety with a base width of just 1.7 m. It would still stand with a base width of only 1.2 m, as shown in Figure 3.1b, and to the modern engineer looks impossibly slender. This is approximately the proportion of the test walls constructed at Bath (Mundell et al. 2009), which showed negligible movement due to backfill alone and could take concentrated loading at the ground surface in excess of 10 tonnes.

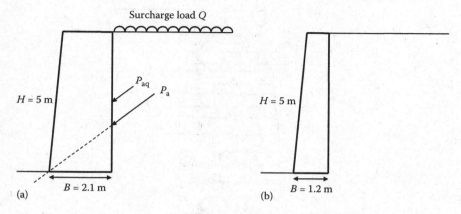

Figure 3.1 The effect of full friction on the back of the wall. (a) With a base width of 2.1 m, the resultant from the earth pressure due to the soil acts within the base of the wall, so produces no overturning moment about the toe. The equivalent force for the earth pressure due to surcharge loading acts higher, and would produce a moment about the toe. (b) Without the surcharge, a 5-m high wall with a base width of only 1.2 m would be just stable.

3.4 OVERTURNING BEHAVIOUR

Figure 3.2 represents a substantial wall built in the French style, with continuous courses of stone from the front to the back of the wall, and no use of small fill stones. Earth pressure acting on the back of the structure could result in sliding or overturning, as described in Chapter 2. If the wall begins to tip over, it will tend to separate into a portion that moves, and a portion that remains resting on the foundation. This separation will naturally occur on a boundary that passes between the stones, which have no tensile connection with each other. Figure 3.2 shows a number of locations for this boundary, but the wall will look slightly different at any other cross-section because the stones have different dimensions and are arranged to overlap each other. It is therefore reasonable to treat this boundary as an inclined plane, as shown by the dotted line. This plane passes through the toe of the wall and is inclined at angle ψ to the horizontal. Earth pressure acting on the part of the wall below the line does not contribute to the overturning moment, only the pressure acting above the line. The overturning moment will therefore be reduced compared with that experienced by an intact gravity wall, but then so will the resisting moment, as the part below the line does not have to be lifted for the failure to occur.

An approximate assessment of this effect might be made for the wall shown. If we take this to be the 5 m high wall we analysed in Chapter 2, then its base width is about 2.2 m, and the width at the crest is 1.6 m. The plane reaches the back of the wall approximately 1.3 m above the base.

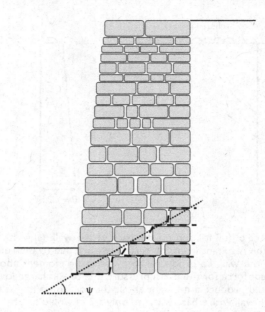

Figure 3.2 Lines of separation within a drystone retaining wall. The action of the earth pressure on the back of the wall could cause the wall to slide or tilt, separating along the dashed lines shown and leaving some blocks on the ground.

Overturning of the portion above the plane will be compared with overturning of the entire wall, were the lower courses to be cemented together sufficiently well for this to happen.

This is investigated in Table 3.1, using a unit weight $\gamma = 15$ kN/m³ for the drystone retaining wall. There is an important difference between these calculations and those of Chapter 2: $\delta = \varphi'$; that is, full friction is mobilised on the back of the wall.

Table 3.1 Overturning analysis

ψ	0 (intact wall)	10	20	27	30
Weight of wall (kN/m)	142.5	142.5	142.5	142.5	142.5
Weight above plane (kN/m)	142.5	136.1	129.3	124.0	121.5
Resisting moment from weight above plane (kNm/m)	177	167.6	157.6	149.9	146.3
Earth pressure P_{ah} (kN/m)	45.7	38.9	32.2	27.5	25.4
Acting moment of P_{ah} about toe (kNm/m)	76.2	74.9	70.9	66.4	63.9
Earth pressure P_{av} (kN/m)	32.0	27.2	22.6	19.3	17.8
Restoring moment of P_{av} about toe (kNm/m)	70.4	59.9	49.6	42.4	39.2
$F_{overturning}$	3.25	3.04	2.92	2.90	2.90

The value of $\psi = 27°$ is chosen because when the calculations are examined with higher precision, this is the approximate value that gives the lowest factor of safety, because as ψ increases so the restoring moment from the weight of the wall above the plane decreases, but so does the disturbing moment from the earth pressure. The exact value of ψ that gives the lowest factor of safety against overturning will vary according to the parameters involved, but the possible values of ψ are limited by the configuration of the stones within the wall. The yield analysis (see Section 4.5) takes this into account in the way in which the wall is 'homogenised' for the purposes of the analysis. However, the simple limit equilibrium analysis presented earlier captures all the essential features of this. In summary, to obtain a sufficiently accurate factor of safety against overturning, it is essential to take into account the actual behaviour of a drystone retaining wall, which depends on the stones it is built from and the way in which they are assembled. However, the range of values presented in Table 3.1 shows that the actual value of factor of safety is not very sensitive to ψ, for which it may be reasonable to just use a worst case value rather than attempting to find a representative value for a particular wall. If the width of the wall is reduced by 0.2 m, the critical value increases to about 30°, while a base width of only 1.6 m results in a critical value of about 38°. As the wall becomes narrower, the range of feasible values is likely to decrease, as there will be fewer stones between front and back, and so fewer possible boundaries on which the wall might separate.

3.5 WALL DEFORMATIONS AND THEIR IMPLICATIONS

If the wall is relatively slender and vulnerable to overturning, the particular nature of the construction becomes important. One aspect of this is discussed in Section 3.4, but there is a broader question of how easily the wall can bend or shear in response to the earth pressure acting on it. This is a consideration for any earth retaining structure, but the particular nature of drystone construction allows parts of the wall to separate from each other, reducing the resistance.

Deformation in bending is shown in Figure 3.3a. In most engineering materials, bending results in a compression on the side away from the pressure and an extension on the side facing the pressure. So in the case of a retaining wall, the face would be shortened and the back lengthened. However, this will happen only if a material compresses as easily as it stretches, and this is not the case for drystone construction. Because the stones are usually strong and stiff in relation to the loads they are carrying, very little compression can take place, so the face of the wall will not shorten significantly. On the other hand, the stones at the back of the wall are simply sitting on top of each other, and so can be lifted off each other. Significant bending deformation can therefore take place only if the overall vertical

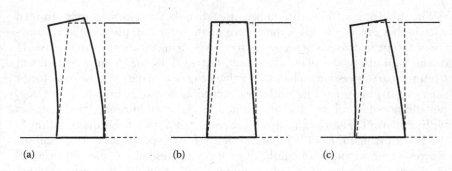

Figure 3.3 Wall deformations. (a) Bending deformation. (b) Shear deformation. (c) Combined bending and shear.

stress within the back of the wall has reduced to zero. If the back of the wall is tending to lift up, then the full frictional resistance will be mobilised to resist that movement, so stones will tend not to lift, until the entire wall fails in overturning. The assessment of whether this can happen then becomes very similar to the assessment described in Section 2.4.2 and Figure 2.13. The position of the resultant force may be considered at any level within the structure, in a similar way as was done for the base of the structure.

If the resultant force at a level within the wall lies in front of the middle third, this does not necessarily imply that the blocks at the back of the wall would lift up, merely that they will not be carrying a vertical load; the earth pressure on their backs would be pressing them against the blocks in front of them, and the friction on those blocks would be supporting the weight of the blocks at the back of the wall. This would introduce a shear stress on a vertical plane within the wall – this always occurs when anything bends. For the structure to deform in pure bending requires a shear connection between the blocks at the front and the blocks at the back of the wall. This may be provided by through-stones, or by having good overlaps as shown in Figure 3.2. Well-packed fill between front and back facing stones is unlikely to have the same effect – the wall will deform in shear.

Shear deformation is illustrated in Figure 3.3b. In one sense this is an easier type of deformation than bending for a wall made of rigid blocks. If the blocks make up layers, then shear deformation only requires one layer to slide on the layer below. On the other hand, drystone walls are usually made of rocks with rough surfaces and good frictional resistance, which makes this type of deformation difficult to achieve. Walls made of horizontally laid slate, with smooth surfaces with only moderate frictional resistance, could be liable to this type of deformation. Nevertheless, because this mode does not require blocks to actually lift up, a modest amount of shear deformation may take place as loads are applied to a wall, and in the full-scale experiments at Bath, a few millimetres of shear deformation occurred as the backfill was placed, arising from very small movements of stone on stone as the friction between the stones quickly attained the value required to maintain equilibrium.

A combination of shear and bending, as shown in Figure 3.3c, is much more likely to occur in a drystone wall. This is because even though stones may not slide over each other easily, nor be lifted up easily, they may be able to rotate. The construction of the wall should make this difficult – a fully bonded construction will help, as will good use of through-stones and tightly packed, strong, angular fill.

Figure 3.4 shows how such an effect may arise in a fully bonded wall. It should be borne in mind that this is only a schematic representation, and that three-dimensional effects that cannot be shown here make the real behaviour more complex. However, this shows how the accumulated rotations and displacements of individual blocks can result in an overall shearing and bending of the structure. Individual blocks tend to rotate because they are being pushed at the back and restrained at the base. Blocks that in this cross-sectional view are wide in relation to their height (i.e., have a high aspect ratio) will not rotate easily. The blocks that experience the greatest rotational forces are those at the toe of the structure, and if these blocks have a low aspect ratio (so they tend towards a square or round cross-section), then they could rotate and overturn, taking the rest of the wall with them. It is therefore not good practice to lay stones along the face of a wall, though because of the three-dimensional nature of the wall, infrequent stones laid in this way are unlikely to lead to a collapse.

The behaviour of a wall that has a front and back face and packed 'fill' in between depends on whether or not through-stones are used. The situation

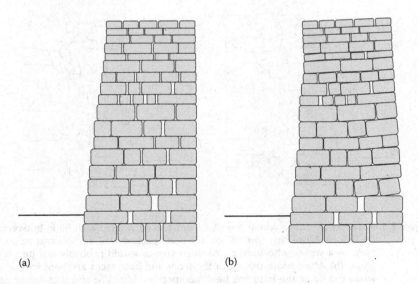

(a)   (b)

Figure 3.4 Rotation of stones within a fully bonded wall. (a) Before and (b) after.

without through-stones is shown in Figure 3.5. The front and back faces are to an extent free to behave independently, with bending deformations arising from rotation of the stones with lower aspect ratio, and the fill material compressing and being rearranged in response to this. The earth pressure exerts a bending moment on the entire structure, but for the structure to respond as one, requires a shear stiffness within the fill which is just not there. So instead of behaving as a deep cantilever with high bending stiffness and strength, the front face and back face respond separately, providing very much less resistance.

Through-stones restrain the front face and the back face from moving apart as the fill settles, so helping to maintain its tight packing, but they also restrict the rotation of the stones immediately above and below them. By ensuring that the weight of the wall and the fill is transferred into the front and back face, they also help to prevent sliding. The result is that the wall below each through stone is very much more rigid, but the through-stones also help the wall to behave as a single cantilever, with greatly increased bending stiffness.

The walls shown in Figures 3.4 through 3.6 are relatively wide, having proportions that an engineer might expect to produce a good factor of safety. The behaviour changes if the walls are more slender, which can work perfectly well and was normal for older walls. Figure 3.7 shows a wall with proportions that would lead to acceptable factors of safety with

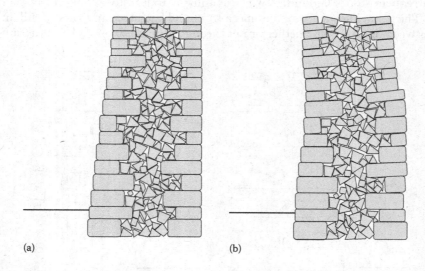

(a)   (b)

Figure 3.5 Rotation of stones within a wall – front and back face, with fill in between. (a) Before. Note that this fill has not been packed well, for illustrative purposes – a waller who used no through-stones would probably not be very good! (b) After. Note that both the front and back faces are bent forward, while the fill at the base has been compressed laterally, and that higher up the wall has settled.

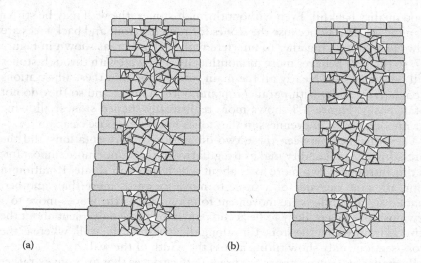

(a)   (b)

Figure 3.6 Rotation of stones within a wall – front and back face, with through-stones. (a) Before. Note that the through-stones would be at intervals along the length of the wall as well as up its height, so this two-dimensional representation has inevitable shortcomings, but it nevertheless can illustrate the constraints the through-stones impose on the deformation of the structure. (b) After. There is hardly any change in response to the application of earth pressure. The through-stones not only tie the front and back faces together; they also increase the pressure on the facing stones, making it much less likely that they will move.

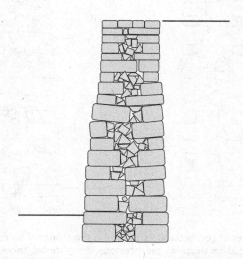

Figure 3.7 A slender wall without through-stones. Having less fill between the front and back faces enables some direct interaction to occur between the larger stones, so this wall may be stiffer than the wider wall shown in Figure 3.5.

good quality backfill. Even without through-stones, the wall may be stiffer than the wider wall because the stones forming the front and back faces are now close enough together to interfere with each other, as shown in Figure 3.7, so the wall behaves more monolithically. The wall with through-stones will be stiffer still. Nearly all the main stones shown in these illustrations are relatively wide, with parallel top and bottom faces, and so they do not rotate easily. Figure 3.8 shows more realistically shaped stones, allowing the importance of unevenness in the stones to be seen more clearly.

The difference between these two-dimensional representations and the three-dimensional reality makes irregularity in the stones more important, as this introduces two more axes about which they can rotate. Rotation in plan, allowing one side of a stone to move forwards more than another, enables significantly more movement to take place as the stones move to a new position; this is likely to be accompanied by some movement about the third axis, in which the stone tilts along the length of the wall, whereas the cross-sections only show tilting across the width of the wall.

If smaller, narrower stones are used, with surfaces that are convex rather than flat, then they can rotate much more easily and larger deformations will result. The amount by which the wall deforms becomes particularly sensitive to the skill and care that has been exercised in placing the stones. The use of pinnings (small blocks or wedges of stone) to stabilise the main stones can become almost inevitable if the builders are concerned to produce an even face to the structure. Because of this sensitivity to the finest details of the construction of a highly variable material, deformations are

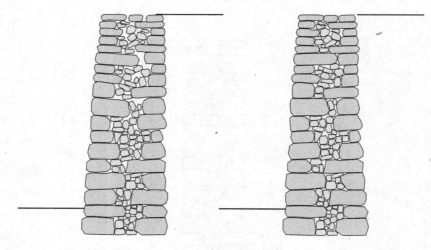

Figure 3.8 The effects of geometrical variations in the stones. This cross-section is less schematic than those in the previous illustrations, showing how the same stones can alter their positions to fit together in slightly different ways as the wall deforms in response to the load on its back. The more loose the original construction, the more it will deform.

likely to be uneven. Nevertheless, skilled builders can produce a wall that might deform by less than 1 in 1000 as it is loaded; this was shown in the first, highest quality test wall built at Bath which was monitored to millimetre accuracy.

3.6 BULGING

Drystone walls often display pronounced bulges in their faces, most typically centred at about a third of their height. Some are the first stages of a collapse, while others probably form soon after construction and remain unchanged. Some observers will say simply that 'bulging is a three-dimensional problem', but in fact long stretches of wall can show a continuous bulge, so this statement is simplistic. It is therefore essential to consider what generates bulging that might be represented on a two-dimensional cross-section, before going on to consider three-dimensional bulges, which may exist over very short lengths of wall and could be a different kind of phenomenon.

3.6.1 Two-dimensional bulging

In considering bulging of any kind, it is important to take into account that the wall has changed from a presumably flat faced condition into a bulged condition, and then stopped moving. It therefore seems self-evident that the wall was not stable in its initial condition, and became stable in its bulged condition. The only plausible reason for questioning this would be if the bulging was a slow deformation that took place while a particularly severe load was applied, which then stopped when the load was removed. If this were the case, one would have to admire the ductility of the structure that could allow such deformation to take place while continuing to support the load, and wonder why the deformations did not accelerate; it is much more plausible that the deformed structure supports the load better, as a stretching spring ceases to stretch when it is in equilibrium with the load hung from it.

A stable two-dimensional bulge can develop because of the interaction between the backfill and the wall. The downwards component on the back of the wall due to friction can result in the back of the wall being compressed vertically more than the front of the wall over much of its height; this can be seen by considering the position of the resultant force in Figure 3.1a. This causes the top of the wall to tilt back a little, which reduces the earth pressure acting on it; this tilting is enabled by the lower part of the wall tilting forwards under the action of the higher earth pressure at this greater depth. The improved stability of the deformed wall was demonstrated by Mundell et al. (2009), using one of the methods of analysis that will be described in Chapter 4. This was discussed further by McCombie et al. (2012).

3.6.2 Three-dimensional bulging

At its simplest level, a three-dimensional bulge can be just a two-dimensional bulge over a limited length of wall that may be weaker than adjacent sections. Three-dimensional bulges seem to be more common than two-dimensional bulges, which is hardly surprising given the inevitable variability within any length of wall. They are often much more pronounced, while still being stable. However, a bulge that develops some time after a wall was built is likely to be a response to either a change in loading, whether externally applied or through water pressure changes, or to a deterioration in materials over many years. In such cases, although the bulge has clearly led to a new equilibrium being reached, the changes that led to its formation are likely to continue and could push the wall to the point where it can no longer stand. This condition is likely to be precipitated by a single stone moving slightly too far, so it tips over. Sometimes this results in a loss of a small part of the outer face of the wall, while the inner face continues to resist the earth pressure, and the rest of the face arches over where the missing material has fallen out. An apparently stable bulge could be on the point of reaching this condition and be highly dangerous; there is likely to be a void within the wall as the face has become separated from the stone behind it,

Figure 3.9 A well-developed bulge in a test wall. This bulge went on to collapse when the load plate on the backfill was pushed further.

and the facing stone could burst at any moment. This extreme behaviour has been observed in working walls, but was also seen in the test walls at Bath (Figure 3.9), which had been constructed with instrumentation so that these details of the behaviour could be confirmed.

3.7 TENSILE STRENGTH

The way in which a three-dimensional bulge is supported is dependent on the form of construction of the wall. In most cases it is dependent upon the tensile strength of the wall (McCombie et al. 2012). Most people's initial reaction to this would be, The stones are just laid on top of each other, with nothing to stick them together, so how can they have tensile strength? To this one might respond, A natural fibre rope is made of lots of short strands, which are not stuck together, so how is it possible to make a rope as long as you like? The answer is friction.

In a rope, short fibres are twisted together into a yarn, and as they try to untwist, they press against each other, so that they cannot be pulled apart unless the resulting friction is overcome. These yarns are twisted together in the opposite direction to form strands, and once again, the strands press together as they try to untwist. Three strands are then twisted in the opposite direction again to form a rope. The torsion in each component makes each press so hard against the other that a small rope made of individual short fibres can have a high tensile strength – a 12 mm diameter manila rope will support over a tonne.

In a drystone retaining wall, the blocks of stone can be simply lifted off each other, but unless every single block was individually held up, then its weight, and the weight of all the stones resting on it, is resting on the stones below. It is then impossible for a stone to move sideways without overcoming the friction resulting from this pressure. This is shown in Figure 3.10. Tensile strength can be developed along the length of the wall face provided that the wall has good bonding; a running joint is a

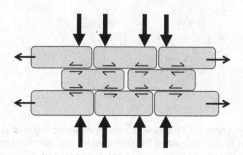

Figure 3.10 The development of tensile strength. Friction between the stones allows tensile forces to be transmitted from stone to stone.

continuous vertical gap with no stones crossing it and so will disrupt the tensile strength. This strength is available in both the front face and the back face of the wall, provided that the same attention to avoid running joints has been given to the hidden back of the wall as to the visible face. This tensile strength can be called into play if there is a localised load at the back of the wall (Figure 3.11); the stones at the face can be pushed outwards, but only to a limited extent because the loaded section has a tensile connection to the adjacent sections, which help to support it. The wall is now acting in a three-dimensional manner, and the face will bulge outwards adjacent to the concentrated loading. If the bulge develops large deformations, then the tensile strength along the length of the face will result in the face acting in catenary along its length, holding back the localised pressure like a net.

The stones themselves are likely to be very stiff, so once they have been displaced by such a load, they will stay in their displaced positions even after the load has been removed. This bulging may only be slight, and may be undetectable, because the amount of movement needed to generate the frictional resistance may be only very small. Because the resistance is

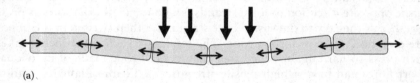

(a)

(b)

Figure 3.11 Catenary action. (a) This view is looking down on a line of stones in the face of a wall. Tensile strength along the face enables the stones to resist localised lateral pressure, distributing the load to adjacent sections of the wall. (b) This view is looking along the top of a test wall that was built straight, and is now bulged outward, so catenary action helps to redistribute the concentrated load.

frictional, it will continue to be available as long as the stones remain in contact with a compression pushing them together.

This type of behaviour is described as plastic, or ductile. The permanent deformation is its main characteristic. In a drystone wall that has been taken to the limit by a short-term concentrated load, such as a heavily loaded wheel too close to the wall, this deformation may be considerable without the wall going on to collapse. A secondary characteristic of plastic deformation is that it absorbs energy; the applied forces move through a distance, while the stones slide over each other providing fictional resistance, and so work is done on the stones. If the load comes from a vehicle impact, for example, then the energy from the rapidly slowing vehicle is expended in the permanent deformation of the wall.

3.8 VERTICALLY ORIENTATED STONES AND BENDING RESISTANCE

Slate and shale in drystone construction is often placed in a wall so that the pieces of stone lie on their edges, rather than on their bedding planes, as is normally done. An example is shown in Figure 3.12. The edges of the stone are usually rough, while the faces are smooth, so this construction provides good frictional resistance on very uneven horizontal surfaces. It can be very quick to build once the stones have been roughly sorted, and the stones may all extend through the full thickness of the wall. Because all the stones are about the same shape they can be quickly placed, on their edges, resting against each other. Often stacks of horizontally laid stones are made at the end of a working section to provide something for the first stones to rest

Figure 3.12 Vertically orientated construction. This wall in West Somerset was built using long pieces of shale placed on their edge, rather than laid flat. This is because their surfaces are comparatively smooth, and they do not develop strong frictional resistance, whereas their edges are rough. Many of the stones extend through the full thickness of the wall.

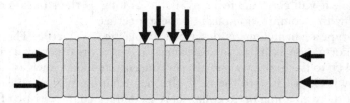

Figure 3.13 Bending resistance. This view is looking down on a line of stones that have been laid on their edges, as shown in Figure 3.12. They have been forced together by wedging action, so that they cannot bend to the deformed shape shown in Figure 3.11.

against. The final stones in a section may be jammed in, so that there are no open gaps between adjacent stones. This means that the wall cannot bend out of plan, as shown in Figure 3.11, and the entire wall becomes very much stiffer, as shown in Figure 3.13.

This bending resistance means that a wall constructed in this style can distribute a concentrated load along its length, but like a beam or slab rather than a net in catenary. However, the mechanism is a little more complex than in a conventional beam or slab, because these rely on tensile strength in their lower parts, whereas the stones are compressed against each other during the construction process, and become further compressed if the wall deforms, so that the wall is acting more like a prestressed concrete beam.

3.9 CONSTRUCTION STYLES USING ROUNDED STONES

In some locations walls are made of stones which are much more rounded, and do not have obvious parallel or even flat faces on which they can be laid. Building with such stones can be challenging, and a common style uses large stones resting with curved faces against each other, but held in place by smaller round stones; only friction between the stones prevents them from rotating and precipitating a collapse (Figure 3.14). Although this type of construction can work, it does not have the tolerance of distortion of walls built with flatter stones, as only a small displacement can break the contact between adjacent stones that is preventing them from rotating.

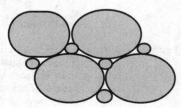

Figure 3.14 Construction with round stones. Large round stones are placed against each other, and prevented from rolling by smaller stones fitted in between them.

3.10 THE EFFECTS OF A TIGHTLY CONSTRUCTED FACE

The aspect of a wall's construction which is most apparent to the client or their agent is the aesthetic of the facing. This has led to construction practices in which a great deal of effort is made to produce a tightly packed face, with the visible surfaces of the stones aligned to that face, to present a flat surface. This may be detrimental to both short-term and long-term stability of the structure.

Figure 3.15 shows the front and back faces of a granite test wall. The front face is made to show tightly fitting joints, but the consequence is that the back face is extremely uneven. This means that the front part of the wall has a higher density than the back of the wall. For maximum resistance to overturning, weight at the back of the wall is more helpful – if there is more weight at the front than at the back, it makes overturning more likely. Hence, the aesthetic requirement is reducing the engineering performance. To achieve a flat face, it is much more likely that pinnings (small wedges of stone) are used to raise the back of a stone. These smaller stones deteriorate more quickly than the larger stones, especially if they have been made by breaking up a weak stone. Good practice requires that only strong stone is broken up to produce pinnings, but if the result is being judged by superficial appearance, then the long-term durability of the structure may be neglected. The consequence of disintegration of the pinnings will be a loss of shape of the wall, as the stones will tend to fall

Figure 3.15 Front and back faces. These two photographs show the front and back face of the same wall. The difference in density of the construction is immediately apparent.

backwards. Near the base of the wall, some stones may actually lean further forwards, due to the greater lateral forces, reducing the stability and possibly precipitating collapse.

3.11 SUMMARY

Drystone walls make very effective gravity retaining walls, their rough backs and permeability leading to optimum performance. The fact that they are made of individual stones resting on each other gives them ductility, an ability to tolerate localised overloading or reduction of support, but also permits modes of deformation and failure that are unique to this form of construction. To a large extent, good construction practice will result in structures that maintain their integrity up to the maximum possible loading for their geometry and weight distribution. Their special characteristics and behaviour are well understood, and Chapter 4 presents methods of analysis and design.

Chapter 4

Analysis and design

4.1 ENSURING SAFETY

A key aspect of engineering design is ensuring that what gets built is not just able to do the job, but can also do it with a margin of safety. This usually means that it will be more expensive, use more materials and take longer to build than if it were only just adequate. Because of this, there must be a clear understanding of why the margin of safety is provided and how large it needs to be. A modern approach to safety is to consider the degree of uncertainty in all the parameters that go into a design, and then either design for the worst credible values or apply a standard partial factor to each design value that has itself been determined in a standard way. This standardised approach makes it much less likely that the engineer will give any real consideration to the actual range of uncertainty in the project being designed. This is made more difficult by the fact that traditional single factors of safety, by which resisting forces must exceed acting forces, were often developed over years of experience to control deformations, rather than to prevent a structural failure. Committees writing books of rules would often aim to ensure that the 'partial factors' approach produced a similar result to the traditional approach, which might make no sense at all in terms of the stated philosophy. For example, a factor of 1.3 or 1.5 on the unit weight of the soil is common, even though it would be almost inconceivable that the unit weight could be this much greater than the design value, even if major mistakes were made at every stage of design and construction.

It is therefore important that when developing modern approaches to designing and assessing long established technologies, the definitions of safety should be based on a proper understanding of the mechanisms that might lead to a loss of function, the variability of the materials and the significance of deformations.

4.2 THE DISTINCTION BETWEEN ANALYSIS AND DESIGN

The design of a drystone retaining wall is the process of making decisions about dimensions, materials and form of construction, having first clarified exactly what function the proposed structure is intended to perform. Analysis in a broad sense means assessing an existing or proposed construction from every relevant point of view. This would certainly cover an engineering analysis of its stability, and is likely to include a consideration of the durability of its materials, but could also include an assessment of its aesthetic qualities and even of its suitability as a refuge for flora and fauna.

Analysis of an existing structure has a focus on what is actually there, whereas in design a wide range of possible options may seem to be available at first, and the analysis is directed at something that might be made. A design process might ideally be a sequence of rational decisions, perhaps including a mathematical deduction of geometry and material properties that will do the job. It might alternatively be a case of proposing geometry and materials from the point of view of construction, aesthetics and durability, and then checking that it has satisfactory stability by a mathematical analysis.

The mathematical analysis of a drystone wall is therefore only part of the design process, and part of the analytical process, but some analysis might actually lead to design decisions directly. For example, consideration of sliding stability might lead to a decision about the width of the structure. Other analyses may only be useful for giving some insight into the behaviour of a structure, but that insight might be generalised to whole classes of structure. It is always clear that the analysis is considering a representation of a real structure, and hence the term 'modelling' is frequently used.

Substantial studies of drystone wall construction and performance have been carried out over the last 25 years, and these studies are ongoing, with the aim of guiding the maintenance, repair and new construction of these structures. Three main methods have been used for modelling drystone retaining walls:

1. Limit equilibrium method (LEM)
2. Yield design method (YDM)
3. Distinct element method (DEM)

Each method has advantages as well as disadvantages. The first two are particularly useful for generating design decisions, while DEM can give useful insights in the analysis of somewhat idealised structures.

4.3 THE DISTINCT ELEMENT METHOD

This numerical method was developed to analyse the deformation of jointed rock in rock mechanics (Cundall 1971). In the field of drystone construction, it was first used by Walker and Dickens (1995) to simulate the behaviour of the free-standing and retaining walls of Great Zimbabwe. It was also used by Harkness et al. (2000) and by Claxton et al. (2005) to model the well-known four drystone wall tests by Burgoyne (1853).

The principle of the DEM is to set up and to solve the equations of motion for the elements. According to the DEM, the system to be analysed consists of discrete elements, which may be rigid or deformable; deformable elements are discretised into triangular subelements, but elements are commonly spherical, or fixed assemblies of spheres to represent more complex forms. The model may be two-dimensional or three-dimensional, though the latter takes far greater computing power. The discrete elements touch at contact points that may change and transmit forces between them at these points. If the forces acting on an element are not in equilibrium a displacement will result, and if moments are not in equilibrium a rotation will result. The analysis proceeds in a series of time steps, chosen to be small enough so that in a single step an element can interact only with its immediate neighbours. For each time step, a force-displacement law is used to determine the contact forces, while applying Newton's second law using the out-of-equilibrium forces and moments defines the instantaneous acceleration of the element, which is integrated to give the velocity. Once the relative velocity between the contact points is known, the relative displacements must be calculated and new contact forces deduced. This cycle is repeated until equilibrium is achieved and movements cease. At this stage, the structure being analysed may or may not be standing: it is possible for the elements to be rearranged completely. If the structure remains standing, then the equilibrium deformations and stresses at the contact points may be obtained. If not, then the movement of the individual elements during the analysis illustrates the failure mechanism. The question then naturally arises as to when the time stepping process begins. Commonly the analysis traces a construction process, with layers of material being added, and the analysis run through to equilibrium. However, Walker et al. (2007) reported a much faster approach of defining the full geometry from the outset, then progressively applying gravity to the model. This approach was feasible for back-analysing structures that were known to have stood, and would work for a structure that can reach a stable equilibrium.

Figure 4.1 shows an example of the distinct element model established by Claxton et al. (2005) using universal distinct element code (UDEC) for wall D in the series of tests carried out by Burgoyne (1853). In this example, the stone, the soil and the bedrock were all considered as Mohr–Coulomb elastic/plastic materials. The cross-sections of the wall, the backfill and the rock foundation were divided into meshes of discrete elements. The backfill soil was treated as deformable while wall blocks were defined as rigid to reduce the running time. Meshes used by Harkness et al. (2000) were about two times denser, and all elements were defined as deformable. The resulting calculation time was quite long using the computers available at

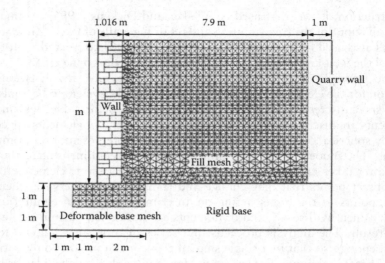

Figure 4.1 **Distinct element model by Claxton et al. (2005). From Claxton, M. et al.,**
ASCE Journal of Geotechnical and Geoenvironmental Engineering 131(3): 381–389,
2005.

the time. It took Harkness et al. (2000) 7 days to run an analysis using an
RS6000 workstation. The calculations of Claxton et al. (2005) took 60–80
minutes using a more modern Pentium II. Although the more powerful
computer contributed to the substantial time saving, it is clear that increasing the number of elements considerably increases the time required for the
calculation.

The DEM is useful for exploring aspects of drystone behaviour, and for
investigating the sensitivity to variations of the input parameters and the
geometry. However, it is not useable for routine design because of its complexity. As with finite element analysis, it is easy to produce a result with
very impressive graphics, but it is very easy for that result to be wrong,
and depth of understanding of the problem and of the analytical method
is needed to ensure that results are good. It is also very time-consuming
for routine work. Besides the material properties such as the unit weight
of stone and the internal friction angle of the wall, the DEM also asks for
knowledge of joint stiffness (normal stiffness and shear stiffness), which
is not easy to determine. In the absence of actual data on stiffness parameters, as was the case for modelling Burgoyne's tests, research on stiffness
properties for rock interfaces may be consulted. There remains inevitable
uncertainty, and as the results obtained may be quite sensitive to this
parameter (Walker et al. 2007), the method cannot be very reliable as a
predictive tool, even though it is useful as an investigative tool. Harkness et
al. (2000) and Claxton et al. (2005) compared the results of drystone wall

behaviour using the DEM and limit equilibrium and found out that the results could be in agreement with the experimental observations presented by Burgoyne. That is, even though the geotechnical parameters were not given by Burgoyne, the differences in the reported behaviour between the four test walls in relation to their geometries could be reproduced using either the DEM or limit equilibrium assessments.

4.4 LIMIT EQUILIBRIUM ANALYSIS

Limit equilibrium analysis has been used for a long time in the design of gravity retaining walls and is based on comparison of stabilising and destabilising actions. In a simple case such as an unreinforced concrete retaining wall, only an external stability is considered. That is, the wall acts as a single monolithic body, and the failure surface is defined as the contact surface between the foundation and the structure. In contrast, for drystone retaining walls there is also the possibility of internal instability with the failure surface passing through the wall; this was, for example, referred to by Harkness et al. (2000) while comparing the results of analysis using DEM and limit equilibrium. Constable (1874) introduced the same idea, carrying out reduced scale experiments using blocks of pine as bricks and oats as backfill. These experiments showed that scaled walls did not overturn in their entirety, but the failure surface made an angle of about 45° with the base. The consequences of this observation will be developed in the description of the limit equilibrium model.

The use of limit equilibrium analysis for drystone retaining walls was studied by Villemus et al. (2007) and then Mundell et al. (2009). Villemus et al. (2007) developed calculations considering that the wall was monolithic, whereas Mundell et al. (2009) presented a computer program that treated the wall as a series of stacked layers to enable investigation of the position of the line of thrust.

4.4.1 Monolithic wall analysis

Figure 4.2 represents the calculation model used by Villemus et al. (2007) for the internal stability of drystone retaining walls. Two modes of failure were considered: sliding and overturning. In both cases, the structure is separated into two monoliths by a plane at an angle ψ to the horizontal. The value of ψ is determined by consideration of the physical characteristics of the structure being analysed. For failure by sliding, the recommended values are 0 if the wall is built from cut stone and 0.2 radian (12°) for rough stone. The authors did not give a specific value for ψ for overturning, but suggested that it should lie between 0 and 12°. The limit will in fact be given by the slenderness of the blocks, that is, how far they extend back from the face in relation to their height. The implication is that stones below this

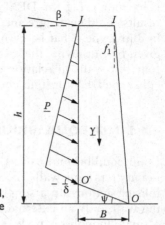

Figure 4.2 Model of limit equilibrium – monolithic wall, used by Villemus (2007), and used to draw the charts of the Appendix.

plane will not be lifted up as the wall above it overturns. The lower the blocks are in relation to their width, the lower this angle will be. While the lower monolith does not contribute to the overturning resistance, the pressure on the back of it does not contribute to the overturning force, and it is not immediately obvious which of these factors would be critical. The yield design analysis described in the text that follows considers the possible values for ψ explicitly, but for the purpose of comparison an angle of 1 vertical to two horizontal (i.e. 26.6°) will be taken to be representative for the limit equilibrium calculations. The same considerations apply to both methods.

The analysis requires a value for the friction angle at the wall-backfill interface (δ), which was not considered by Villemus et al. (2007) in their analysis of hydraulically loaded experimental structures, which could not apply any friction on the internal face of the wall. However, in the general case of earth backfill, based on the work of Powrie (1996) and Colas et al. (2008), δ may be set equal to φ_s – the backfill friction angle. This value was also used by Mundell et al. (2009), and gave results that corresponded very well with full-scale tests pursued to overturning failure (see also Chapter 3).

4.4.1.1 Wall stability against sliding

The wall is considered to be potentially stable against sliding if the safety factor of wall stability against sliding is not less than 1. See the discussion in Chapter 2 about an appropriate safety coefficient.

For ψ = 0:

$$F_{slid} = \frac{V \tan \varphi}{H} > 1 \tag{4.1}$$

where φ represents the friction angle of the blocks sliding on each other, or of the blocks sliding on the foundation material, whichever is the lower, and is determined by shearbox tests. V and H are, respectively, the vertical and horizontal components of the resultant of external forces applied to the failure part OIJO'.

Solving the inequality (4.1) will give us B^{slid} – the minimal base thickness required to assure the internal stability of the wall against sliding, while replacing 1 with an appropriate value of F_{slid} for design, usually 1.5, will result in a satisfactory construction.

For $\psi > 0$, V and H are replaced by the forces acting normal to and along the failure plane.

4.4.1.2 Wall stability against overturning

The wall stability against overturning is potentially ensured if the resisting moment (M_r) is greater than or at least equal to the overturning moment (M_{ov}). In other words, the overturning safety factor

$$F_{\text{ov}} = \frac{M_r}{M_{\text{ov}}} > 1 \tag{4.2}$$

By satisfying this inequality, we can find out B^{ov} – the minimal base thickness required to ensure stability against overturning. As with F_{slid}, for the design of a new structure we would seek a margin of safety, and typically F_{ov} must be 1.5 or 2.0; 1.5 is used here for the design of drystone retaining walls.

In the end, we have the ultimate base thickness required, defined as

$$B^{\text{ult}} = \max\{B^{\text{slid}}, B^{\text{ov}}\} \tag{4.3}$$

4.4.2 Multiblock wall analysis

Mundell et al. (2009) developed a program using the Delphi development environment to analyse the stability of walls, which was tested against both models and full-scale tests carried out at the University of Bath (McCombie et al. 2012). This program was created to investigate the position of the 'line of thrust' within the wall (Cooper 1986) and how it changes in response to changes in a wall's geometry and loading. A wall is stable provided the line of thrust remains within the width of the wall; this is equivalent to the analysis used in arch bridges (Heyman 1966, 1988). The aim of the program was to give some insights into wall behaviour very quickly, by comparison with the time-consuming complexity of distinct element modelling as described earlier. The wall is considered to be composed of a series of stacked blocks with horizontal upper and lower surfaces. Each block

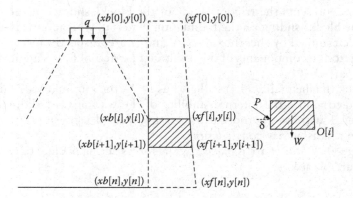

Figure 4.3 Model of limit equilibrium – multiblock analysis.

extends from the front to the back of the wall and represents a complete course of stone within the real structure. It is identified by the coordinates of its four vertices (Figure 4.3), from which its area and centroid are determined. The geometry and position of these blocks thus determines the overall geometry of the wall. The program allows this geometry to be altered by a mouse-click on the cross-section shown on the computer monitor, or by entering new values into the table of data. The new positions of the resultant forces at each block interface are shown virtually instantaneously, together with the line of thrust. Provided that the line of thrust lies within the structure, then overturning will not occur.

The applied loads to each block are composed of

- Block weight (W): This is calculated by multiplying the area and the unit weight of the material. The load is applied at the centroid of the block.
- Backfill pressure (P): This is represented by a force acting at an angle of δ with the normal of the internal face, placed at a height determined by the difference in pressure between the top and bottom of the block.
- Surface loading (q): The influence of a load applied to a limited area of the ground behind the top of the wall is determined by assuming that the load spreads out over an area that increases with depth by a ratio of (1 horizontal : 2 vertical) in all directions, but is limited by the position of the back of the wall. The surface load application is taken into consideration only when this load spread touches the wall.
- Load transmitted from the block above (zero for the topmost block).

The calculation begins from the block at the top and continues to the lowest one. To evaluate the wall stability, the program checks three possible failure modes: overturning, sliding and block rotation. The wall is no longer stable when the sliding or overturning forces exceed the resisting ones.

Although based on the same theory of limit equilibrium, Mundell's approach differs from that of Villemus, as it considers different mechanisms. The reason comes from their different aims: Villemus wanted to build a model to design new drystone retaining walls whereas Mundell aimed at assessing the stability of existing walls. Villemus assessed the stability of the part of the wall above a single failure line whereas Mundell checked the stability at the level of each course. The failure line used by Villemus's case was in fact a zigzag line passing through different layers of stones, whereas Mundell's was a straight horizontal line separating two courses.

Besides the two familiar failure modes of sliding and overturning, Mundell's program also considered the rotation of an individual block at the front of the wall. If the resultant force comes close to the face, the horizontal thrust as well as vertical load can rest on a single stone. If this stone is high compared with how far it extends back into the wall (i.e. it is slender), then it can rotate forwards, precipitating failure of the rest of the structure. This is a simple matter to check – the use of limit equilibrium analysis requires the engineer to consider which failure mechanisms might occur, but it requires knowledge of the geometry of the blocks of stone used in the construction.

As noted in Section 4.4.1, whereas Villemus had no need to consider the influence of the value of δ (soil/stone interface angle), Mundell, like Colas (see Section 4.5), proposed to take $\delta = \varphi_s$ (internal friction angle of the soil), on the basis of the rough face presented by the drystone construction to the backfill.

Mundell took into account for the first time a surcharge load. Though a strip load would be strictly compatible with the two-dimensional analysis, the implemented approach was easily extended to a square or rectangular applied load, to give equivalent values that could be used. However, if the loading was localised, the wall experiencing this loading would be restrained by adjacent unloaded sections of wall. This restraint is actually dependent on the quality of the construction, a fact confirmed by the full-scale tests carried out at Bath.

4.5 YIELD DESIGN ANALYSIS

In a general case, the yield design is used to determine the ultimate load that a structure can sustain. Two approaches can be used: an interior (static) approach, which is based on statically admissible stress fields and gives the lower bound of the ultimate load; or an exterior (kinematic) approach, which is based on the virtual work theorem with the study of kinematically admissible virtual velocity fields and gives the upper bound of the ultimate load. Colas et al. (2008, 2010a,b) chose a kinematic approach

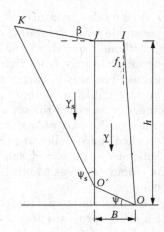

Figure 4.4 Drystone retaining wall modelling.

in combination with the homogenisation theory developed for periodic masonry (de Buhan and de Felice 1997) to model drystone retaining walls.

Yield design theory requires three kinds of parameters: geometry of the system, loading mode and resistance of the constituent material. In the studied case, the problem geometry is defined by a height H, thickness at the bottom B, front batter α and backfill height H_b (Figure 4.4). As would be expected, there is a strong degree of similarity between Figures 4.2 and 4.4.

The loadings considered in the study are the respective unit weights, γ_{DW} and γ_s, of the wall and its backfill soil. The wall was approximated as built from rigid regular cut stone blocks with dimensions a and b, so that it could be considered as periodic. The joints are assumed to have a purely frictional Mohr–Coulomb shear criterion, depending only on the block friction angle φ_b.

It is then possible to consider the cell represented in Figure 4.5 to implement the homogenisation process in the framework of yield design theory.

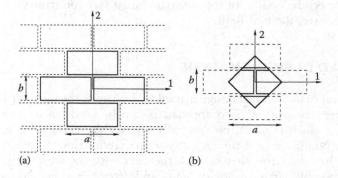

(a)   (b)

Figure 4.5 Cell of the periodic masonry. (a) Actual wall. (b) Unit cell.

The macroscopic strength domain G^{hom}, describing the set of macroscopic stress states $\underline{\underline{\Sigma}}$ such that there exists a stress field $\underline{\underline{\sigma}}$ defined over the cell (C) and verifying the following conditions, is written

$$\underline{\underline{\Sigma}} = \underline{\underline{\sigma}}(\underline{x}) = \frac{1}{V_{cell}} \int_C \sigma(\underline{x}) dV \tag{4.4}$$

$$\text{div } \sigma(\underline{x}) = 0 \tag{4.5}$$

with $\underline{\underline{\sigma}} \cdot \underline{n}(\underline{x})$ antiperiodic, with $n(\underline{x})$ being the unit normal oriented outwards from the cell C, and $\sigma(\underline{x}) \in G(\underline{x})$ whatever $\underline{x} \in C$, $G(\underline{x})$ characterising the strength capacities of the constituent materials.

In this approach, the kinematic definition of G^{hom} will be used, which can be obtained through the dualisation of the static definition by means of the principle of virtual work. One considers any virtual velocity field of the form

$$\underline{v}(\underline{x}) = \underline{\underline{F}}\underline{x} + \underline{u}(\underline{x}) \tag{4.6}$$

with $\underline{\underline{F}}$ any second-order tensor and $\underline{u}$ a periodic velocity field. The strain rate field $\underline{\underline{d}}$ can be written as

$$\underline{\underline{d}}(\underline{x}) = \underline{\underline{D}} + \underline{\underline{\delta}}(\underline{x}) \tag{4.7}$$

where $\underline{\underline{D}}$ is the symmetric part of $\underline{\underline{F}}$ and $\underline{\underline{\delta}}$ is the strain rate field associated with $\underline{u}$.

The principle of virtual work leads to:

$$\int_C \underline{\underline{\sigma}} : \underline{\underline{d}} \, dV = \int_{\partial C} \underline{\underline{\sigma}} \cdot \underline{n} \cdot \underline{v} \, dS \quad V\underline{\underline{\sigma}} \text{ Statically Admissible}, V\underline{v} \tag{4.8}$$

By replacing $\underline{v}$ by its expression, and because u is periodic and $\underline{\underline{\sigma}} \cdot \underline{n}$ is antiperiodic, it follows that

$$\int_C \underline{\underline{\sigma}} : \underline{\underline{d}} \, dV = \int_C \underline{\underline{\sigma}} \, dV : \underline{\underline{F}} \tag{4.9}$$

so

$$\langle \underline{\underline{\sigma}} : \underline{\underline{d}} \rangle = \underline{\underline{\Sigma}} : \underline{\underline{F}} = \underline{\underline{\Sigma}} : \underline{\underline{D}} \tag{4.10}$$

where

$$\langle . \rangle = \frac{1}{V_{cell}} \int_C . \, dV \tag{4.11}$$

Introducing

$$\pi\left(\underline{\underline{d}}\right) = \max_{\underline{\underline{\sigma}}}\left\{\underline{\underline{\sigma}} : \underline{\underline{d}}, \underline{\underline{\sigma}} \in G\left(\underline{x}\right)\right\} \text{ and } \pi^{hom}\left(\underline{\underline{D}}\right) = \max_{\underline{\underline{\Sigma}}}\left\{\underline{\underline{\Sigma}} : \underline{\underline{D}}, \underline{\underline{\Sigma}} \in G^{hom}\right\} \tag{4.12}$$

which are the respective support functions of the convex domain $G(\underline{x})$ and G^{hom}. The static definition of G^{hom} leads to

$$\pi^{hom}\left(\underline{\underline{D}}\right) \leqslant \pi\left(\underline{\underline{d}}\right) \quad \forall \underline{u} \text{ periodic} \tag{4.13}$$

Then it may be proven that

$$\pi^{hom}\left(\underline{\underline{D}}\right) = \min_{\underline{u}}\left\langle \pi\left(\underline{\underline{d}}\right)\right\rangle \tag{4.14}$$

which constitutes the kinematic definition of G^{hom}. We may then consider a particular velocity field $\underline{v}$ defined as

$$\underline{v}(\underline{x}) = \underline{v}^i \tag{4.15}$$

where $\underline{v}^i$ is the velocity of any block C^i (see Figure 4.6).

The obtained strength domain is represented in Figure 4.7 in the field $(\Sigma_{11}, \Sigma_{22}, \Sigma_{12})$. m is the slenderness ratio (a/b) of the blocks and $f = \tan \varphi_b$, with φ_b being the friction angle between the blocks.

The support function of the strength domain (Figure 4.7) of the homogenised material is obtained as

$$\pi^{hom}(\underline{n}, \langle \underline{v} \rangle) = 0 \tag{4.16}$$

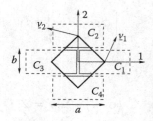

Figure 4.6 Velocity field solution in the cell.

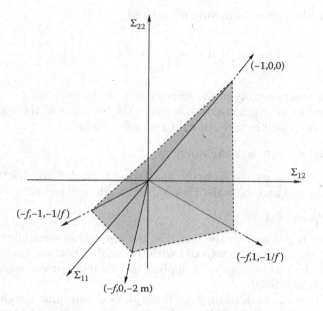

Figure 4.7 Homogenised strength domain of drystone retaining wall.

with the conditions

$$-n_1 v_1 \leq 0$$
$$\tan(\phi_b) n_1 v_1 \leq 2m n_2 v_2$$
$$|n_1 v_2 + n_2 v_1| \leq -\tan(\phi_b) n_1 v_1 + n_2 v_2 / \tan(\phi_b)$$

(4.17)

else

$$\pi^{\text{hom}}(\underline{n}, (\underline{v})) = +\infty$$

(4.18)

where $\underline{v}$ is the discontinuity of the virtual velocity field in the wall and v_i ($i = 1, 2$) are the components of the velocity $\underline{v}$ of the cell constitutive blocks.

The soil is considered to be a Mohr–Coulomb material, depending on its cohesion C_s and friction angle φ_s characterised by

$$\pi(\underline{\underline{d}}_s) = \frac{C_s}{\tan(\phi_s)} \, \text{tr} \, \underline{\underline{d}}_s \quad \text{if} \quad \text{tr} \, \underline{\underline{d}}_s \geq (|d_{s1}| + |d_{s2}|) \sin(\phi_s)$$

(4.19)

where $\underline{d}_s$ is the rate deformation tensor and

$$\pi\left(\underline{n}_s, \underline{v}_s\right) = \frac{C}{\tan\left(\phi_s\right)} \underline{v}_s \cdot \underline{n}_s \text{ if } \underline{v}_s \cdot \underline{n}_s \geq \left|\underline{v}_s\right| \sin\left(\phi_s\right) \tag{4.20}$$

with $\underline{v}_s$ any discontinuity of the velocity field in the soil.

The interface strength criterion between the back face of the wall and the backfill is described by a frictional Coulomb interface:

$$\pi(\underline{n}, \underline{\Delta v}) = 0 \text{ if } \underline{n} \cdot \underline{\Delta v} \geq |\underline{\Delta v}| \sin(\delta) \tag{4.21}$$

where $\underline{\Delta v}$ represents the velocity discontinuity and δ the friction angle between the soil and the wall. This friction angle will be taken as

$$\delta = \min \{\phi_s, \phi_b\} \tag{4.22}$$

This choice is justified by the fact that the optimal discontinuity line will always be localised in the medium with the smaller friction angle.

Second, the yield design was applied to calculate an estimation of the ultimate backfill height.

Noting that the back of the wall is not smooth but quite rough, the friction angle at the wall-backfill interface δ was set equal to the backfill friction angle φ_s. Knowing all necessary parameters, the possible ultimate backfill height was calculated in the framework of the kinematic approach of the yield design theory (Salençon 1983, 2013). This theory is based on the principle of virtual work combined with the knowledge of the strength criterion and leads to the inequality between the work of external actions (W^e) and the maximum resisting work (W^{mr}) for all kinematically admissible virtual velocity fields as a necessary condition of stability:

$$W^e \leqslant W^{mr} \; \forall \; \underline{V} \; KA \tag{4.23}$$

Two different mechanisms of failure will be shown here: translation of the wall and soil (Figure 4.8a) and wall rotation and soil shearing (Figure

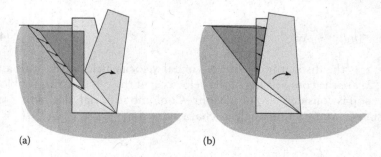

(a)   (b)

Figure 4.8 Studied failure mechanisms. (a) Translation of the wall and soil and (b) wall rotation and soil shearing.

4.8b). The smaller result of the two cases was taken as the final result. These last failure mechanisms were verified by 2D scale-down tests using Schneebeli rods to simulate backfill soil in two dimensions (Colas et al. 2010a).

It should be noted that the backfill height is considered different from the height of the wall, and the wall stability is evaluated in relation to the soil height that the wall could support. The stability problem could be characterised by the following nondimensional factor depending on nondimensional parameters:

$$\frac{H_b}{H} = F\left(\frac{B}{H}, \alpha, \frac{\gamma_{DW}}{\gamma_s}, \phi_s, \phi_b\right) \tag{4.24}$$

The factor of safety in this case can be defined as

$$F = \frac{H_b^{real}}{H_b^{limit}} \tag{4.25}$$

in which H_b^{real} is the actual height of the backfill soil, while H_b^{limit} is the maximum height of the backfill soil required to ensure stability.

Although the yield design theory is more complicated than the limit equilibrium, it has been considered to give better results than the approach of Villemus, which depended on the value of the angle ψ between the horizontal and the failure line through the wall, which can be estimated by measurement, whereas in the yield design this angle is calculated in the optimisation process. However, this process depends on the geometry of the construction in the same way as in the limit equilibrium approach of Villemus, because that failure plane steps up through the courses of masonry in exactly the same way as it can in the homogenisation. The difference is that homogenisation implicitly allows steeper angles provided that they also step through the structure in a similar way, whereas Villemus implicitly checked for just $\psi = 0$ and for the first stepping value. A thorough limit equilibrium check would, as a matter of course, consider these mechanisms also, but observations of test walls confirm the assumption that those considered by Villemus would normally be critical. In all cases, a conscious decision must be made to consider steeper values of ψ, and the actual possible values depend on the geometry of the stone used in the construction.

4.6 DESIGN CHARTS

Design charts are graphs that summarise results of calculations using either the limit equilibrium analysis or yield design theory as presented in the preceding sections. The DEM is not considered here as it is not used for design. Design charts can be used to provide an initial indication of the expected geometry in an initial design of drystone retaining structures. Charts from CAPEB et al. (2008) are reproduced in the Appendix of this book. The expertise of such masons is required to ensure compliance with good practice, to produce final constructions that allow the engineering assumption of monolithic behaviour of the walls in 2D. This was verified by experiments (Villemus et al. 2007; Colas et al. 2008; Mundell et al. 2009).

These charts were established using the limit equilibrium theory for the monolithic wall which was presented in Section 4.4.1. The friction on the back of the wall was assumed as $\delta = \varphi_s$. A safety factor (F_{sli}, F_{ov}) of 1.5 was applied, which is usually used on gravity retaining walls in France.

4.6.1 Utilisation of design charts

The guide provides 18 charts in total, corresponding to 2 kinds of stone, 3 values of backfill slope (β) and 3 different values of external batter of wall face (f_1) as follows:

1. Materials: limestone, schist
2. Backfill slope: 0°, 10°, 20°
3. External batter: 0%, 10%, 20%

Figure 4.9 gives an example of the design charts that are found in the guide, showing the case of walls in schist with external batter of 10% and a backfill slope of 10°. The x-axis represents the backfill friction angle (φ_s)

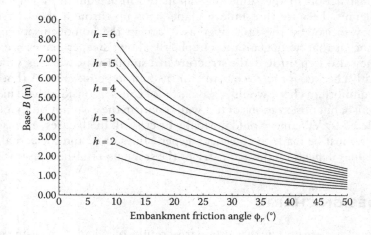

Figure 4.9 Design chart for schist drystone retaining walls with β = 10° and f_1 = 10%.

measured in degrees while the y-axis represents the base thickness of the wall (B) measured in metres. Ten curves are given corresponding to 10 different wall heights varying from 1.5 m to 6 m. Therefore, once he or she knows φ_s, the engineer can preset the wall height (h) and then consult the chart to find the base width of the wall. This is the minimal value to ensure that the wall is stable. For example, for a 2.5 m high schist wall retaining a backfill with friction angle 30°, the minimal base width required so that the wall remains stable is 1.4 m, allowing for the safety factor. The thickness at the coping could also be calculated if necessary, based on the three parameters: h, f_1 and B, 1.15 m in this case.

It should be noted that only walls with vertical internal faces ($f_2 = 0$) and horizontal courses are considered. For other cases of β and f_1 that cannot be found in the list above, there are two ways to solve them: we could either use the method of linear interpolation or simply take the closest value of the reference parameters, while erring on the safe side.

4.6.2 Graphical comparison between the results of limit equilibrium and yield design

The results of limit equilibrium analysis and yield design modelling are compared for a drystone schist wall of 2.5 m height (Figure 4.10). A safety factor of unity is chosen in both approaches for the purpose of the comparison. From the graph it may be seen that the two approaches are very close, with a maximum difference in base width of only 5 cm at an angle of friction of 50°. This difference is small by comparison with the required factor of safety and indicates that either method can be used with confidence in design.

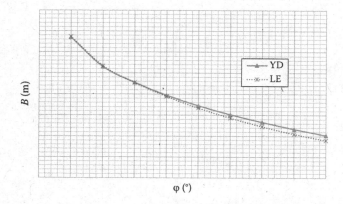

Figure 4.10 Comparison between results of yield design and limit equilibrium ($h = 2.5$ m; $\beta = 10°$; $f_1 = 10\%$).

4.7 SUMMARY OF ANALYTICAL METHODS

Studies on drystone structures are still ongoing to increase the depth of understanding, especially with respect to a wider range of construction styles. In France and in the United Kingdom, where the transport infrastructures depends on a large number of drystone retaining walls, this research has economic and environmental importance. Replacement of efficiently designed drystone structures with concrete alternatives has serious environmental and aesthetic impact and may be significantly more expensive. Comparisons between the different methods described here, and with full-scale tests carried out in France and in the United Kingdom (Villemus et al. 2007; Colas et al. 2008; Mundell et al. 2009), indicate that the behaviour of drystone retaining walls is understood in detail and can be predicted. The DEM allows parametric investigations and back analysis, while LEM and YDM are suitable for routine engineering design. Design charts have been presented for a range of cases. Therefore, when drystone structures need to be replaced, engineers can be confident in replacing them with newly designed drystone structures that will meet current engineering standards. These structures will be efficient and sustainable and will sit well in their environment.

Chapter 5

Construction

5.1 BUILDING IN DRYSTONE

Drystone retaining walls have the unusual characteristic that they are built of just a single material – stone. The construction of a wall requires the study of the environment in which it is to be built, before making the best choice of dimensions and construction methods, which depends on knowledge of the stone, tools and instruments used in the construction.

5.1.1 Environment

The construction or restoration of a wall must take into account its environment. The geology, geography and landscape must all be understood if the best decisions are to be made. Study of the context can give precise understanding of the ground above and below the wall. Observations of existing works can give an indication not only of the particular local construction methods, but also of any potential problems such as weak or compressible surface layers, or difficulties arising from water circulation. Dimensions of existing walls may be determined from where they have been cut through, and this will inform decisions about the dimensions for repair or replacement – it is unwise to take existing practice as a direct guide for construction or repair.

5.1.2 Material

It is the local geology, allied to the needs of economic development such as farming and transport routes, that has both led to the need for drystone walls and determined the variations that are found in different localities. It is important from the aspect of landscape and cultural heritage as well as aesthetics that walls are made of local stone, of the same geological origin as neighbouring walls. Furthermore, the use of local stone obtained close to the site reduces the environmental impact of the construction through reducing energy used in transportation, as seen in Chapter 1.

5.1.2.1 Geological considerations

From the geological aspect, stones fall into one of three families:

1. *Sedimentary rocks – principally limestone and sandstone*: These are formed by the deposition and solidification of sediments of organic or mineral origin. These rocks are very variable because their genesis depends on many factors – the nature of the sediments, their mode of transportation, the place of deposition and the way in which they are turned into rock (diagenesis). In general they consist of stratified deposits in layered beds. Calcareous (limestone) deposits have a characteristic bedding and jointing that produces individual stones of a rectilinear form, while sandstone gives beds of varying thickness, which can erode to give more rounded forms. Sedimentary rocks have variable durability.

2. *Igneous rocks – principally granite and basalt*: These rocks form through the crystallisation of magma, liquid rock. These rocks are not formed in layered deposits and are typically found in more or less rounded forms, according to the erosion they have experienced from wind and water. The rubble stone that is used in construction may be obtained by breaking up the stone by impact, by large hammers or by picks.

3. *Metamorphic rocks – principally schist or gneiss*: These are formed by the transformation of other rocks by heat and/or pressure. Schist and gneiss both have their origin in clay sediments, which in their unaltered state are not suitable for wall building. Shale is clay that has just begun to undergo metamorphism and is sometimes used in drystone construction if no more suitable material is available, though even then it may be described as 'slate', which is harder and stronger, but has surfaces on which friction is low, so special techniques are required to make a strong wall. The next stage of metamorphosis is schist, which when unweathered can provide a strong material. Schist tends to break into layers easily, though the layers are typically thinner and of less even dimensions than is typically found with limestone. Gneiss can come in various forms and is hard and strong; most commonly it has to be used like granite, but it can sometimes delaminate into layers when it weathers.

The stones used in the construction of drystone walls must be of good quality. In particular, they must have good resistance to compression, which is to say, they must have sufficient compressive strength to resist the forces they must carry. In general, this corresponds to the hardness and density of the rock. For stones that are formed of many layers, such as schist, it is necessary to ensure that the stones do not break down into layers too easily, and that the layers are tight together. It is also necessary to ensure that the rocks do not break down when frozen, and those with the best resistance to water penetration will be chosen.

5.1.2.2 The supply of stone

The vast majority of rock used in drystone construction has come either from the clearance of stones from fields, or direct extraction of rock from the construction site itself. Nowadays, most stone comes from quarry extraction, although it may sometimes be obtained from other sources such as demolition or excavation.

5.2 BUILDING A DRYSTONE RETAINING WALL

The stability of a drystone wall depends on the quality of its implementation as much on its proper sizing. This requires compliance with the established rules of the art, and on the skill of the drystone wallers. Although a simple stack of stones could in theory provide the function of a drystone retaining wall, the durability and ductility that make this form of construction so resilient in the face of variations in loading and ground conditions are entirely dependent on the construction quality, while other aspects of good practice provide some tolerance of variability in the construction materials.

The waller must know the various arrangements of stone and elements of a wall to make the best decisions about the arrangement, sorting, selection and placement of the stones (Section 5.2.2.1). Building a wall begins with the site preparation and the organisation of the work (Section 5.2.2.2). Once the site is ready, construction of the wall according to the rules of the art can begin (Section 5.2.2.3).

5.2.1 The different elements of a drystone wall

Within French practice, each type of arrangement of the stones, each part of the wall has its name; each stone and each side of each stone has a name according to its position. Accordingly, clear communication and understanding requires these names to be defined.

Note that each region has its own labels to denote the materials or their implementation, in addition to the variations between languages. The etymological diversity of these terms makes it futile to attempt to list them all; instead we will try to employ here the generic terms most commonly used, offering English approximations where necessary, supported by illustration and explanation to ensure clarity of communication.

5.2.1.1 Arrangement of stone

The way in which the stones are arranged within a wall, and so transmit forces from stone to stone, is called the 'opus'. Three factors come into play in the choice of the arrangement of stone for a given wall: its function, its environment and the type of stone available. An aesthetic choice of the customer or builder can also come into play.

The opus is most apparent in the outward appearance of the wall, but the internal organisation of the stones, as well as those visible at the face, must be carefully implemented according to the selected opus.

5.2.1.1.1 Geological influences

Geological characteristics, shape and design of the material will dictate the first choice of arrangement of stone. The two major categories of arrangement of stone are coursed masonry (opus assisé) and opus incertum.

1. *Coursed masonry*: Stones from more or less regular thinly-layered strata are generally used as blocks, with clear top and bottom beds derived from that layering, or bedding. In the case of sedimentary rocks, these surfaces on which the stone naturally splits are called the bedding planes, and they were horizontal when the sediments were deposited, before they turned into rock. Regional metamorphic rocks split on cleavage planes that are defined by their schistosity, and these may not be parallel to any original bedding planes, but as with the sedimentary rocks, they will split into layers on planes that are orthogonal (perpendicular) to the direction of maximum stress they experienced as they developed their present form, and this means that they will have their greatest strength and stiffness in a wall if they are placed with these planes horizontal (Figure 5.1), so that they are again carrying their greatest loads in a direction orthogonal to their bedding or schistosity. Rocks that do not have either characteristic do not usually break down into neat layers that can be used this easily in masonry, which led to the next opus, described in the text that follows.
2. *Opus incertum*: Some stones come in a form that does not allow them to be placed in layered beds without being resized; this makes the arrangement of stone 'uncertain'. (Figure 5.2) The stones are set in relation to their form, interlocking against each other and using their own geometry. Each stone must as far as possible be in contact with adjacent stones on every side, whatever the number of sides. This is the case for granite, some sandstone, basalt rock, as well as certain limestone.

Although these are the major types of arrangement of stone, there are others, and some walls can combine different opus types.

- *Opus cyclopean*: When the walls are built with very large stones, it is called Cyclopean opus – by reference to the work of the mythical giant 'Cyclops' (Figure 5.3). This is so whether or not the stones can be placed in clear layers.
- *Opus clavé* (vertical construction): This is used in areas where it is imperative that the wall is sufficiently free draining to cope with large water forces. These might be flows through the wall, such as in

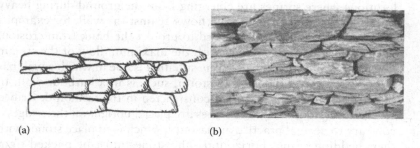

(a) (b)

Figure 5.1 Coursed masonry. (a) Drawing. (b) Example in schist in St Germain de Calberte, France.

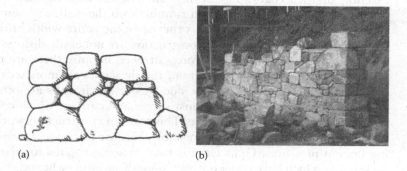

(a) (b)

Figure 5.2 Opus incertum. (a) Drawing. (b) Example in granite, experimental wall building in process at Pont de Monvert, France.

Figure 5.3 Opus Cyclopean, near Le Collet-de-Dèze, France.

locations where springs are emerging from the ground during heavy rain. The forces may come from flows against the wall, for example, if it is constructed on a creek bed to protect the bank from erosion, where it may have to contend with the maximum flow of the stream (the talweg). The most challenging situation is one in which a wall has to resist sea waves driven by a storm, such as in a harbour wall. In these situations walls are often constructed in the fashion described as 'clavade' (Figure 5.4). The stones are placed upright on their edges – contrary to normal practice in masonry, which is to place stones with their bedding planes horizontal. The stones must be packed tight against each other to keep them vertical, but even so, the frequent vertical joints make this form of construction less susceptible to clogging by fine soil and resistant to strong flows through the wall. The stones within the wall, as well as those at the face, should be placed carefully on their edges to assist the flow. The stones at the face will often have their longer dimension running into the wall, and sometimes stones will be chosen so that they span the entire width of the wall. The stones in this form of construction are not easily dislodged by the impact of waves or by strong currents, because there are no large flat horizontal surfaces to act on, and the stones are more secure because they are pressed together side by side. Such walls are often curved to form a vault effect against surges and current, which considerably increases their resistance. This form of construction works well on ground that is relatively unstable, because it is good at resisting uneven pressures. Opus clavé is used in some regions for slates and shales, which have comparatively smooth surfaces, whereas their edges are rough; because their weight is resting on their rough edges, the walls have good resistance to failure on internal sliding surfaces. It is also commonly used to form the coping of a wall (see Section 5.2.4.3).

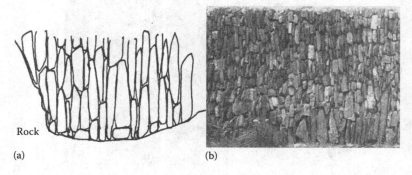

(a)   (b)

Figure 5.4 Opus clavé. (a) Drawing. (b) Example in schist in Boscastle, United Kingdom.

5.2.1.1.2 Social, cultural or artistic influence

The quality of the arrangement of stone can also attest to the technical skill of the builder and the richness of his creative mind. Some arrangements or decorations, or even playfulness in the form of the construction may express cultural traditions or artistic intentions – these also demonstrate the skill of the builder.

- *'Opus quadratum and Roman'*: When the stones are cut 'square', that is, with two right angles to their facing, and are arranged in carefully coursed masonry with excellent bonding, the resulting regular arrangement of stone is called 'opus quadratum' (Figure 5.5). The opus quadratum is built with rectangular stones with horizontal courses that can vary in thickness along their length (Figure 5.5). In the 'Roman opus', each bed is of rigorously constant height. This arrangement can be made with granite, sandstone, limestone and sometimes schist as well. Both of these opus types require substantial work in cutting stone and adjusting its size, especially in comparison with the tradition of using the stones as they are found. A substantial additional skill is required from the waller, and the process takes much longer than methods that focus on the strength of the construction rather than a rigid protocol for its appearance. The observer of a completed wall might suppose the blocks to be rectilinear throughout the thickness of the wall, making them easy to place level and well supported, but these rectangular blocks may only form the outer face of the wall. It is also possible that only the outer face of a stone is rectangular, and the stone may even be supported on wedges to ensure the correct alignment of that face. The skilled waller will nevertheless have built a wall that is structurally sound, even though inordinate effort has gone into providing an appearance to please the imperial masters!

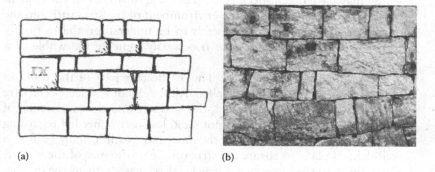

(a)   (b)

Figure 5.5 Roman opus. (a) Drawing. (b) Example in limestone at Balsiegès, France.

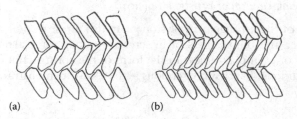

Figure 5.6 Opus spicatum/piscatum. (a) Spicatum. (b) Piscatum.

- *Herringbone pattern and fishbone pattern, "Opus spicatum and pis-catum":* The 'opus spicatum' (meaning like ears of corn, also called laying herringbone) and opus piscatum (referring to the fishbone) are two arrangements already in use in Roman times and still tradition-ally found in some areas (Figure 5.6). The stones are inclined, and the direction of inclination of each course is reversed compared to that of the preceding course. In the case of herringbone pattern, the beds are nested by overlapping the upper and lower ends of the stones, whereas in fishbone pattern, the stones rest on the ends of the stones below without overlap, resulting in a well-defined line of contact. This type of arrangement of stone is achieved with flat stones or oblong pebbles.

5.2.1.2 The parts of a wall

We can divide a drystone wall into parts that are given a specific name (Figure 5.7).

- *The foundation (or base):* This is the first bed of stones on which the wall will stand. The base can either be aligned with the face of the wall or protrude from the facing and would then be described as a footing. Foundation rafts may also be used, in which the stones are set vertically.
- *Exterior facing:* This is the visible part of the wall, for which the waller may have taken special care regarding the finish and alignment of the stones, depending on the environment of the structure and the degree of control required. It needs to be understood that a tightly constructed outer face does not necessarily indicate the whole of a wall is tightly constructed.
- *The back of the retaining wall:* This is the rear part of the wall, on which the thrust of the soil is applied. Although not visible, special care will be given to the arrangement, the wedging and the overlapping of the stones. As the inner face is not seen, less care is needed regarding the alignment of the stones, but the inner face is still usually built to a well-defined plane to ensure the structural performance of the wall. A rough finish to the inner face is beneficial, as it strengthens the interac-tion with the retained soil and enhances the stability of the structure.

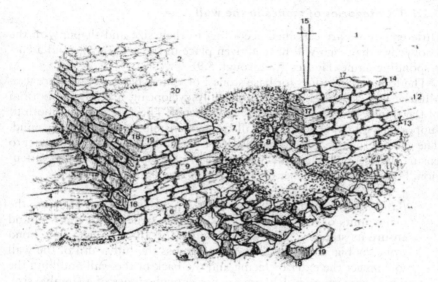

Figure 5.7 The parts of a drystone retaining wall. 1, Terrace cultivation; 2, drystone retaining wall; 3, rubble; 4, rock base; 5, foundation; 6, first course with a greater thickness than the wall, to form a footing; 7, soil excavated prior to building the wall, and replaced as fill; 8, drain; 9, shaped stones; 10, pin stone (for use as a wedge); 11, fill stones; 12, courses; 13, the bed of a course; 14, coping; 15, the angle of batter of the face; 16, sequence of stones forming a corner; 17, through-stone; 18, face of the stone; 19, heel of the stone; 20, terrace; 21, soil; 22, internal facing stone; 23, external facing stone.

- *Internal organisation of the wall*: Except for the through stone (see Section 5.2.1.3), the stones of the exterior facing and those of the back of the wall do not often touch each other, because of the thickness of the walls. The stones of these two sides are linked together by stones of different sizes, which are well pinned and crossed over. These stones create the internal organisation of the wall that is crucial to its structural integrity. Rigorous care must be applied to this internal construction throughout the building process.
- *The covering (or coping)*: This is the last bed of stone placed on the wall. It is made of larger, heavier stones, to hold in place the lighter stones forming the highest courses of the wall and to connect the inner and outer facings. These heavier stones are less easily dislodged by people or animals, and so help prevent degradation of the upper parts of the structure. The stones may be either placed flat, to form a level top to the wall, or on edge. This latter arrangement discourages animals from trying to cross the wall, and can be made with stones that are quite varied in shape, provided that they are of sufficient size. Locally available materials, local practices and the type of wall all influence the way in which the coping is constructed.

5.2.1.3 Categories of stones in the wall

Different stones are classified according to their size and shape, with the knowledge that each will have a given place in the wall, a role and a corresponding name (Figures 5.7 through 5.9).

The foundation stones (or base stones, or course blocks): These are usually the largest stones available, possibly cyclopean, and fill the bottom of the trench in which the wall is constructed. They are intended both to support the total weight of the wall and to transmit it evenly to the ground. The ground they are placed on should be carefully levelled and compact to ensure the best possible contact with it. Their face is often given little attention, because they are generally partially or entirely buried.

- *Building stones (through-stones, tie-stones, boutisses):* They are the major units of masonry that make up the structure of the wall and ensure its stability. The 'through-stone' or 'through' is a long bond stone, as big as possible, stretching across the full width of the wall to connect the exterior facing and the back of the wall and bind the whole construction. If there are not enough stones of a suitable size, then two stones may be laid adjacent to each other to join the two faces together as 'tie-stones', or 'three-quarter throughs', one with a face on the outer face, the other with a face on the inner face, and with as much overlap as possible. These stones are then held together with

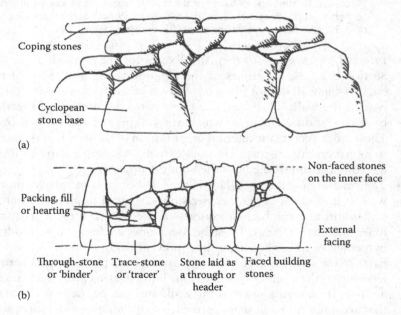

Figure 5.8 Stone arrangement in a wall. (a) Elevation. (b) Horizontal section showing internal organisation.

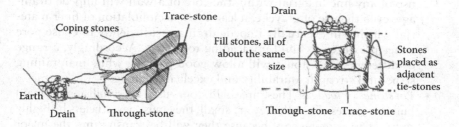

Figure 5.9 The main stones in the wall and the internal organisation.

stones laid above them, resting on both stones; in French, these are called *'boutisse solidaire'* or *'epingle'*. The visible end of these stones forms part of the face of the wall and is usually finished accordingly. From the face, a through may be indistinguishable from the other stones, though the requirement to span the full width of the wall may result in the through being somewhat larger than most other stones.

In contrast to the throughs and tie-stones, the trace-stone (or tracer, or stretcher) is a stone whose longest face is laid parallel to the facing of the wall. It links together two or more stones of the interior or exterior facing. The tracer can be used to stop the beginning of a running joint, and it must be thick enough to be stable without creating a weakness in the wall. However, because the trace-stone does not penetrate into the wall, this can create its own weakness, which should be compensated for with a through-stone or tie-stone placed adjacent to it.

- *Exterior facing stones*: These are of varying size, and have a face that either naturally or after some cutting or trimming is suitable for the external facing of the wall. In French it is commonly described as a *moellon*, which might be best translated as a 'masonry unit', and implies that it has been trimmed to shape. When an 'engineered' appearance is sought, every facing stone may have been trimmed for fit and for external finish, significantly adding to the cost of the time it takes to build the wall, and hence its cost, but possibly having no effect whatsoever on its function.
- *Stones for the back of the wall*: These are stones of similar size to those used for the exterior face, but will not be as finished as the back face of the wall is not visible. In some forms of construction they will not be used.
- *Packing stones or 'hearting' or 'fill'*: These may be of any size, and are usually stone that was not used as facing stone. They should be packed tightly together and placed carefully to ensure that there are no large voids, or the core of the wall could settle leading to distortion or even structural failure. For the same reason, it is important that the stone is of good quality, for if the hearting does not have the same durability as the facing stone, then it will limit the life of the wall. The

use of any fine material within the core of a wall will impede drainage; even the use of gravel can lead to an accumulation of finer material which will eventually impede drainage sufficiently to cause pore water pressures to build up, leading to failure. Accordingly, a range of different-sized stones will allow good packing while maintaining structural strength, durability and excellent drainage.

- *Pin stones or wedges*: These are small stones used for stabilising the main building stones. Though they are small, they must nevertheless be highly resistant to compression, because they will be transmitting the major structural forces through the wall. They may be flat, wedge-shaped or any shape allowing good packing of the stones. Structurally, it is much better to tolerate a little unevenness in the face of the wall than to use many thin wedges, which are likely to deteriorate comparatively quickly.
- *The coping stones (or covering)*: These are the stones that hold together and protect the top of the wall and are chosen according to the chosen method of construction. In the case of a flat covering, the stone must be chosen large enough to connect the two stone facings and heavy enough to be difficult to dislodge. Often, they also prevent penetration into the wall by fine particles carried by surface water, which would clog the gaps between the stones. These stones are also to be relatively flat to ensure a good finished appearance and will also need a presentable edge to form the top of the exterior facing. The back edge should not project significantly beyond the inner face, or it could be dislodged when stood on, or by the wheel of a vehicle, because it is supported only by the backfill and not by the much stiffer drystone construction.

Should a coping of vertical stones be chosen, fairly flat stones will be used that can be cut and sized to be aligned with the inner and outer face, and alongside each other. Some styles use alternating high and low stones to produce a finish that strongly discourages animals, but the stones must still align with the front and back faces. The stones should be placed with their vertical surfaces pressed together, so that one cannot be moved without adjacent stones also moving, thus greatly increasing the strength of the coping and its resistance to impact.

Whichever type of coping is built, if the stones are not wide enough to connect the two facings, they must overlap each other across the width of the wall so that inner and outer faces are tied together and held in place by the weight of the coping.

The stones of the drain behind the wall: These are coarse debris or gravel, often unusable remnants from the dismantling of old walls, or scraps from trimming the building stones. They are used to fill in the back of the structure, to form an additional drain and protect the rear wall facing against the progressive invasion by fine soil. They act as a filter between the backfill and the wall. Depending on soil type, these drains can maximise the protection of the wall, provided it is done correctly.

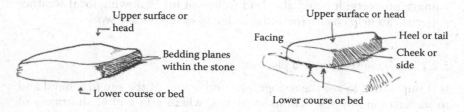

Figure 5.10 The different parts of a stone.

5.2.1.4 The faces of a stone

We distinguish each side of a stone according to its location in the wall (Figure 5.10):

- Lower course or bed, the side that is in contact with the lower bed
- Head or upper bed (*lit d'attente* in French), the side on which the next course will rest
- Facing or face, the visible part of the stone once in the wall
- Cheeks (*joue* in French), the side faces, in contact with adjacent stones along the length of the wall
- Heel or tail, the part of the stone which enters into the wall

5.2.2 Organisation of the construction site

5.2.2.1 Preliminary site preparation

Before beginning work, the drystone builder must reflect on the general organisation of the site. A preliminary analysis must be made of the environment, to identify any peculiarities and difficulties and hence to better organise the tasks and work to be done.

5.2.2.1.1 Information

First, it is necessary to obtain accurate information about the site, its boundaries and any utilities (electricity, gas, water), roads or paths that may cross it. It is also necessary to ask about the weather, not only from general forecasts, but also from local knowledge; drystone construction often takes place in hilly and mountainous regions that can have very localised weather patterns. This knowledge can not only help ensure a proper understanding of potential limitations in the working conditions and time, but also enable adequate preparations to be made in terms of equipment. The weather is also profoundly linked to the risk of ground instability – a major slope failure could not only be a serious impediment of the work and damage property; it could also lead to loss of life. Knowing local weather patterns can help ensure the suitable design of temporary works.

5.2.2.1.2 Site inspection and planning

It is imperative to see the topography and layout of the site first hand and to understand any restrictions of access, whether in width, sharpness of corners, strength of the road or ground or presence of overhead lines or bridges. Access to the site may severely limit the kind of equipment that might be used in the construction – wheelbarrows may have to be used instead of lorries, or mini-excavators instead of larger and more powerful machines. From understanding the access issues, it will be possible to determine the order of operations to perform, the site organisation (earthworks, scaffolding, supplies, work and storage areas) and then the most suitable equipment. Depending on the type of site, the stone volume to be ordered will be calculated. For example, in the case of a restoration, the original stones will be assimilated into the wall and will be complemented by freshly quarried stone; reusing some of the original stones will help the wall to integrate into its environment.

5.2.2.1.3 Signage

The site must be prepared by marking out the access, storage areas and the site of the wall itself by fences or barriers to prevent any risk to passers-by. The work being done should be set out on signs where the public may have access, and appropriate safety notices placed for the attention of those working on or visiting the site.

5.2.2.1.4 Preparation

Safe access routes must be provided and cleared of any obstacles that may block them. It is also necessary to clear the work area of vegetation, and cut down any trees located too close to the work, especially if they are behind where the wall is to be built, as removal of soil and damage to roots could lead to them falling onto the workers. Finally, the transport of stone and scaffolding must be arranged. To ensure greater efficiency and greater security, it will be important to keep the site clean and tidy.

5.2.2.1.5 Management

It is necessary to ensure that everyone involved has a proper knowledge of safe working on construction sites. Only a very low wall (perhaps less than 1 m high) that does not require excavation into a hillside might be suitable for amateur builders. Although this book is intended to provide readers without prior knowledge of civil engineering a good knowledge of drystone retaining walls, it should be apparent by this stage that these are civil engineering structures, and professional civil engineering input may be needed to ensure a safe process and a safe final result, even if the actual building is carried out by wallers with prior experience of retaining wall construction. Anyone using machinery should be properly trained and certified in its use; readers are reminded that operation of machinery on steep slopes, especially transporting materials, is particularly hazardous. Local authorities should at least be consulted, not only to ensure that proper permissions have been obtained, but also to draw from their knowledge of practices and problems particular to the area. Finally, it must be recognised that drystone walling can be heavy and demanding work, and in planning the project it is important that personnel are given sufficient time to do the work well and safely, making due allowance for weather conditions, time taken to get to and from the site, and for rests during the work.

5.2.2.2 Foundation preparation and earthmoving

The first stage of the work is to prepare the foundation on which the wall is to be built, as well as level areas on which the building stone can be placed, and spoil deposited for later reuse. A sufficient area needs to be allocated and clearly marked out to allow the materials to be sorted according to their final place in the construction, and for the stones to be prepared and arranged ready for building.

Then the excavation of the foundation of the wall can begin, ensuring to leave safe and comfortable working space behind the inner face, so that it can be built to a sufficiently high standard. Depending on the purpose of the wall, some or all of the excavated material may be kept for filling behind the completed wall, or in front of it, the remainder being cleared from the site.

It is essential to secure the stability of the ground at the back of the excavation in which the wall is to be built. The nature and stability of the soil, the risks related to weather (freezing and thawing cycles, washing out after heavy rain, softening of the ground through raised groundwater pressures) or the height of the face must all be considered. It is unlikely that anything steeper than the slope shown in Figure 5.11 would be safe in any circumstances; Figure 5.12 shows very basic patterns of formwork for securing potentially unstable soil. The style on the right includes no bracing and serves to do little more than protect the builders from small debris rolling down the hillside.

The danger of working in front of an inadequately supported slope cannot be overemphasised. Most people do not realise that to be buried in soil even just up to waist level can result in death. The debris from crushed tissue overwhelms the kidneys and may lead to death within a few days; for this reason, if this should happen it is imperative that casualties receive prompt medical attention even if they seem unharmed once they have been rescued.

Therefore, even just 1 m of cut needs to be taken seriously and requires advice from a suitably experienced and qualified person – a professional

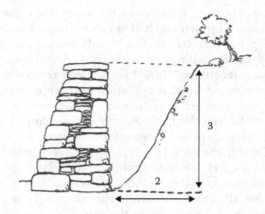

Figure 5.11 Excavated slope.

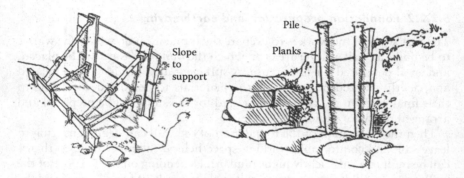

Figure 5.12 Formwork to stabilise the backfill during the construction work.

civil engineer should be able to say whether or not a specialist is needed, but there are situations in which experienced professional wallers can make good decisions, and part of their professionalism is knowing when specialist help is needed.

There are circumstances in which an unsupported excavation 1 m deep can result in movement of an entire hillside, and drystone retaining walls can be very much larger than this. The overall situation must be taken into account when designing the wall, to ensure long-term stability, but the design stage must also consider the safety of the construction process.

In some particularly difficult situations, backfilling will need to be done before the formwork can be removed, and some parts of the formwork must be left in the ground. If this may happen, it becomes important to design the formwork with gaps to allow drainage of groundwater.

Attention must also be given to the drainage of surface water. Gutters should be dug to intercept and lead away any water flowing down a hillside towards the wall, and measures·should be taken to ensure that the foundation excavation is drained. Any water that does wash across the site could pick up soil and debris, causing problems farther downhill. It is therefore helpful to minimise the water flowing across the site, and it may be helpful to cover stockpiles of excavated material with sheeting, both to prevent them becoming saturated and unstable and to prevent material being washed away to produce a muddy outflow from the site.

5.2.2.3 Delivery to site and sorting of materials

Thought must be given to the areas used for sorting and storing the stones and other construction materials. Examples of the kind of arrangement that can make the work easier are shown in Figures 5.13 and 5.14. At the front of the base of the wall, closest to it and ready for use, are the heavy foundation stones. There should be a sufficiently wide passage between the stockpile and the wall to have space to work in comfort and safety. This will enable at least a wheelbarrow to pass and eventually, once the lower part of the wall has been built, there will be space for a scaffold to be installed.

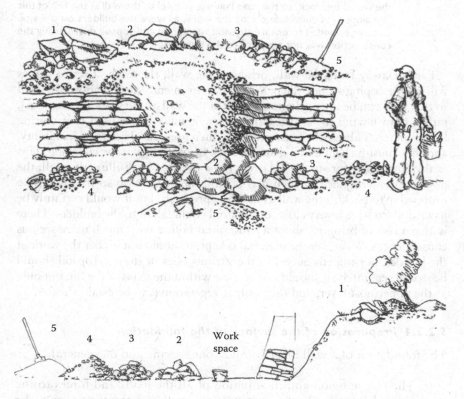

Figure 5.13 Sorting the materials. 1, Covering stones; 2, facing stone; 3, rubble blocks; 4, debris for drainage and filling; 5, backfill and earth.

Figure 5.14 Example of a work in progress of a granite drystone retaining wall in Pont de Montvert, France. This site was a broad roadside verge, offering relatively little space. It may be seen that most of the material in the cut face behind the wall is bedrock, so the road running parallel to the wall at the top of this low slope was quite safe during the work, as were the builders on the site. Care was needed to ensure that the vegetation and topsoil overhanging the construction was not a hazard to the builders.

Farther away from the wall, or above the wall, the farthest set of stones will be the coping stones. They are most conveniently stored above the wall so that they can be lowered onto the top of the wall as the work is completed, rather than having to be lifted. However, it must be realised that storing heavy material above the excavated face increases the chances of instability, and they must be kept a safe distance back from the top of the cut slope. This is also important for stockpiled earth to be used for backfilling the wall; the most convenient place to keep it is behind the excavation so that it can be dropped in to backfill the wall as building proceeds, but it would certainly be hazardous to have it very close, or material could fall onto the builders. There is also a risk of bringing about a substantial failure with much more serious consequences. Wherever the material is kept, it should not hinder the work of the builder, nor impede access to the various piles of stones. Topsoil should be stored separately; it should not be used within the mass of the fill, but only as the uppermost layer, and then only if vegetation is to be established.

5.2.2.4 *Preparation of the surface of the foundation*

The foundation of a wall fills two roles, one specific and one general:

1. The transmission and distribution of all the loads and forces to the ground (principally the weight of the wall and the thrust from the backfill), without any catastrophic collapse.

2. The dimensional stability of the lower part of the structure, without undue settlement and distortion which would dislocate stones to the extent that they would no longer work as intended.

The preparation of the foundation of a retaining wall is literally fundamental to its function, and it is imperative that the material provides good support. According to the geological environment, the wall can be based either on rocks or on soils – a wall on a hillside is more likely to have a bedrock foundation, whereas one close to the valley floor is more likely to have a soil foundation.

5.2.2.4.1 Foundation on a layer of rock

In some sites, particularly in steep areas, the foundation may be *in situ* rock. The foundation must be laid on bedrock that has been cleared of soil and severely weathered material and cut to provide an appropriate sloping surface to best ensure the stability of the wall it is to support. There are four important rules to remember:

1. A friable or degraded rock must be cleared away to reveal stronger underlying material, if at all possible. Sometimes the zone of weathered material may be very deep, in which case a rock foundation is unattainable, and the degraded rock must be treated as if it were soil.
2. The rock must be cut so that its base is perpendicular to the selected batter (Figure 5.15). The slope must not be towards the front of the wall, as this will make sliding easier. Channels may be cut to facilitate the flow of water if necessary.
3. The width of the first course is provided by the prior calculation of the wall design (see Chapter 4). However, if the rock is very hard, and difficult to work, the width of the foundation may be divided into several steps, as shown in Figure 5.15; the stones will provide good support as the construction progresses.

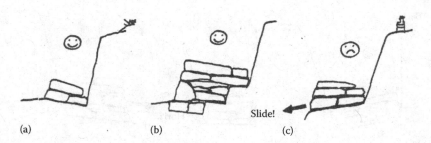

(a)   (b)   (c)

Figure 5.15 Inclination of the foundation on rock. (a) Perpendicular to the batter. (b) Steps. (c) Not to be done.

4. In the case of excessive slope in the longitudinal direction of the wall, the foundations can be formed as a series of horizontal steps, so that the courses of the wall will always be level. This may be essential to prevent longitudinal sliding.

5.2.2.4.2 Foundation on soil

On sites that are less steep, where the substrate is made of thicker layers of earth, the rock substratum is too deep to be reached. It is nevertheless necessary to excavate to a level where the soil is sufficiently firm, whether by manual or mechanical means, according to access and resources available. Care must be taken in the preparation of the foundation because a poorly prepared soil could lead to an overturning failure of the wall, brought about by excessive settlement or bearing failure at the toe of the wall. The following rules must be respected, some of which are of course comparable to those applicable to rocky surfaces (Figure 5.16).

1. The soil must be dug so that the foundations are below the finished floor. This is firstly to find better compacted soil. However, it is also important to ensure that should there be erosion or other excavation in front of the wall, perhaps for cultivation or the making of a paved surface, then the toe will not be undermined. It is also desirable, for most soils, to have the founding level lie below the depth affected by freezing and thawing. The strength of the underlying soil may limit the height of wall that can be built and may dictate that the width of the wall is wider than might be expected. The question of bearing capacity is another matter that requires the input of a specialist; a geotechnical engineer will ensure that appropriate investigations are carried out, and with the understanding of drystone retaining walls communicated in this book, will be able to recommend a safe founding level and wall thickness so that the bearing pressure applied to the soil does not exceed its safe bearing capacity. On a particular site, the engineer may be able to indicate how the soil should be assessed on

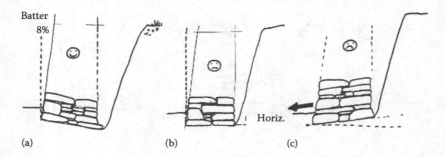

Figure 5.16 Inclination of the foundation on soil. (a) Advised. (b) To avoid. (c) Forbidden.

excavation to enable a decision to be made regarding whether or not the depth achieved is satisfactory.

2. The width of the trench depends on the width required for the wall (calculated during the design; Chapter 4), which as noted in point 1 may be determined by the resulting bearing pressure, as well as by the required size of the wall to retain the earth behind it. It is useful to leave a space in front of the toe of the wall, to facilitate the construction of the foundation layers of stones.

3. Be careful not to loosen the soil in place. If the soil is already loose, it should be compacted or covered by a bed of gravel before receiving the foundation stones. This bed of gravel or stone debris will improve the drainage of the soil under the first course.

4. Once the foundation soil is ready, it should be levelled in the longitudinal direction of the wall. In the case of excessive terrain slope, a series of steps must be formed to prevent the wall sliding downhill and to create level bases from which construction can be done.

5. Transversely, the base should be sloped towards the retained soil, at least perpendicular to the selected batter. A soil foundation will not have the shear strength of the rock used in the construction, and a wall founded on soil is particularly at risk of failure by sliding forwards on its base (Figure 5.16). If the foundation slopes backwards, this helps to resist the sliding, whereas if it were to slope forwards it would make sliding easier, and accordingly this is not permitted.

6. The collapse of a drystone retaining wall on soft soil is often due to the settlement of the front of the stones of the first course. Therefore, the construction of an enlarged base is highly recommended because by giving a better 'foot' to wall, the bearing pressure under the toe is reduced, and hence the wall is less likely to overturn. The foundation may be enlarged throughout its embedded depth, projecting beyond the face of the wall above ground level, to form a footing (Figure 5.17). The projection may ordinarily be 5–10 cm, or more for higher walls or on weaker ground.

Backfill

Figure 5.17 Foundation stones larger than the thickness of the wall, providing a footing (the origin of the term is obvious in this illustration).

5.2.2.5 Dimensional control

5.2.2.5.1 Templates (batter frames) and chalk line

A significant aspect of the appearance of the exterior facing of a drystone wall is its good alignment – the faces of the stones should lie within a plane. This alignment is ensured through the use of a template, or batter frame constructed out of slats, wooden rafters or metal rods. It must withstand the inevitable knocks and jolts that it will suffer during construction, and must not move throughout the process. The templates should be checked regularly (Figure 5.18).

The template allows cords or mason's lines to be stretched between them on their insides. Then stones will be placed behind the line, by a short distance of about 1 cm, so that the stones do not displace the line from its true position, so that there is consistent 'daylight' between line and face.

The line, always taut, will be moved upwards as work proceeds. The more uniform the proximity of the stones is to the line, the more even the face will be. To verify the general alignment, one can from time to time check the plane of the wall visually with respect to the inside of the batter frame.

For better alignment accuracy, particularly when placing large stones, the double cord technique is adopted. A first cord is stretched just above the ground and the second at least 30 cm above. During the construction of the lowest course, the stones should not project beyond the plane defined by the two lines together. After the first course, the cords do not get in the way of the mason in all actions of the construction. The second line can be raised as construction proceeds and continue to serve as a reference for the profile.

Figure 5.18 Template and chalk line.

Figure 5.19 Template with a double line (limestone wall at Hauterives, France).

5.2.2.5.2 Interior facing

For the construction of the interior facing, which because it is not visible does not require a perfect regularity, the builder opts mostly for alignment by eye, but guidance can also be provided by fixing timbers onto the templates so that their insides will define the interior face (Figure 5.19).

5.2.2.5.3 Curved wall

When the wall is on a curving alignment, maintaining the desired profile becomes more challenging. The waller may choose to set a curve by eye, while maintaining the batter and the thickness of the wall – this obviously requires great attention and a degree of mastery to achieve a satisfactory result. The two straight sections can be constructed to begin with and then the connecting curve aligned by the eye based on both sides already built.

But it is also possible to place ranging poles at equal distances, say 1 m, into the ground to define a desired curve. A portable template is then constructed and placed on the section of the wall at each of these locations, to ensure that the stones are placed to the correct profile, while the curve is followed accurately. This technique, though laborious, gives a much more satisfactory result from an aesthetic point of view than the common practice (not recommended) of defining the turn by a series of short straight sections, creating a multitude of angles that break the smoothness of curve.

5.2.3 Construction rules

5.2.3.1 Determination of the batter

It is advisable to implement a batter to the external facing. This batter is the rearward inclination of this face relative to the vertical. It moves the wall's centre of gravity to the inside giving it more resistance to tilting. It also helps ensure that the small movements that take place as the wall takes up its full load, and as the foundation compresses, do not result in a wall face that is actually leaning forwards. Everyone who is aware of the force acting on the back of a wall is instinctively more comfortable with a structure that seems to be leaning back against the facing – this is what one expects a retaining wall to do. Even a face that is actually vertical can seem as if it is leaning forwards, and does not inspire confidence.

The thrust of the backfill is minimal at the top of the wall and greatest at the base. This is why the top of the wall does not need to be as wide as the base. If the courses of stone are laid perpendicular to the batter (Figure 5.20), the wall resistance is improved (see Section 5.2.3.2), and this may be significant for stones that do not have good rough surfaces. This also reduces the amount of stone used and the working time.

This batter can be very small or even zero if the wall is less than 1 m high or if it is not subject to major thrusts. It can reach 25% for very slender walls subject to high stresses. The use of the template made for the desired batter helps the builder to maintain the correct geometry through the full height of the wall during construction. It may also be noted that in some traditions of construction, the batter of the face may vary through the height of the wall, the toe of the wall curving outwards, so that the wall becomes steeper as it gets higher. This produces a wall profile that respects the line of thrust within the wall, resulting from the force it must carry;

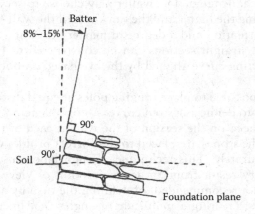

Figure 5.20 Inclinations of the layers perpendicular to the batter.

such a profile comes close to the ideal, because the overturning effect of the backfill pressure does not just increase linearly with depth (Figure 5.21).

On the back of the wall, the inner facing is often built vertical, or stepped, but it should never be further back at the top than it is at the base, or the resulting distortion could cause settlement of the back of the wall and a bursting of the front of the wall. The wall must be stable by itself and not leaning against the slope. As for the front facing, the resistance to sliding will be increased if the stones are inclined towards the back of the wall.

5.2.3.2 General principles of construction

Though field wall construction may be undertaken by individual masons, it is recommended that retaining wall construction is carried out by a team, which facilitates the handling of such large stones and leads to greater safety. For each stone of the wall, the four steps of the construction process may then be carried out to an extent in parallel, as well as in strict sequence.

1. *The choice of the stone*: According to the pattern of construction to be adopted, stones are selected according to their size, shape and strength to determine the places they will occupy in the wall and the roles they will play. For example, it is common to place the larger stones at the bottom of the wall.

   It is important to have a wide choice of stones, allowing the builder to select the most suitable unit for the wall at the stage it is at. It is this choice of stones that ensures the satisfactory performance of the final construction. An experienced waller keeps a visual memory of

Figure 5.21 Kumamoto Castle, Japan: Retaining walls with a curved profile. (From Wikipedia 663highland. Available at https://commons.wikimedia.org/wiki /File:Kumamoto_Castle_02n3200.jpg. Creative Commons License.)

the shapes of different stones in the stock, so that suitable stones for each location can be found quickly.

2. *The preparation of the stone*: Once selected, the waller prepares the stone for the chosen location, thinking of the next stones to be placed. This preparation is always a compromise between raw stone and cut stone, the aim being to minimise the work to be performed on each unit. Depending on the nature of the stone, and the desired final appearance, a good retaining wall can be built with very little shaping of the stone. It is advisable to avoid destabilising the structure of the wall by shaping the stones while resting them on the surface on which they are to be placed.

3. *Placing the stone on the structure*: Each stone where it is placed must bridge a gap between stones underneath it, so intercepting what might otherwise lead to a running joint. This bridging is essential for the ductility (flexible strength) of the structure. It is advisable to ensure that each stone is tight against adjacent stones, to maximise friction between them, facilitating the distribution of stress within the structure.

4. *Wedging*: This is used to improve direct contact between the stone's base and the upper face(s) of the stone(s) of the previous course. Each stone must be carefully wedged, so that it does not rock. The rule is that maximum stability is ensured for each stone, with the isostatic position, also called three-point calibration (Figure 5.22). With three points of support, the stone may seem secure, but when stones are placed above it, it could still move.

Any gaps left underneath the corners of a stone must then be filled with wedges, also called pins. If the stone being used for the wall is generally made up of nearly flat pieces, it may be possible to avoid the use of wedges altogether, and this will always be preferable, but otherwise the need for wedges must be accepted. Filling these gaps increases the density of the wall and is a factor in the quality of a gravity wall. Stones of any size can be used in this operation, but they must be strong enough to withstand

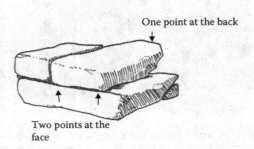

One point at the back

Two points at the face

Figure 5.22 Stabilisation of the stone by three contact points.

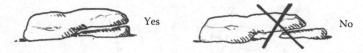

Yes   No

Figure 5.23 The placing of wedges.

the compressive forces to which they will be subjected without breaking. On the other hand, it is necessary to ensure that the wedges do not project beyond the sides or back of the stone being wedged, so as not to impede the placing of the next stones.

For some flat and fine stones, where the risk of failure by compression or bending is high, the contact surface with the lower bed should be maximised by the careful placement of wedges to multiply the number of points where the stone is supported (Figure 5.23). This ensures good stability of the stone. It must be kept in mind that each nonstabilised stone threatens the stability of those that will be placed above.

Each stone, however small it may be, has a role to play and is an essential element in the cohesion of the whole structure.

5.2.4 Raising the wall

5.2.4.1 Foundations

It must be stated here that there is a fundamental principle of this construction method: drystone retaining walls are built either on the rock or on the ground – the foundations of this type of wall must always be drystone. It is neither helpful nor stronger to use concrete or reinforced concrete in the foundations, though this is seen too often. Concrete foundations create a barrier to the natural flow of water, typically channelling the water to the point in the wall that is most vulnerable to pressure and sliding. In addition, they will prove far less flexible than a drystone base: they tend to break instead of adapting to the small movements of the soil. These breaks will pass through the wall as fracture lines, instead of the movements distributing in a diffuse way through the entire structure.

The laying of the foundation stone calls for special care: it is important to lay the stones and place them as far as possible in a through-stone position, that is, with their greater length across the wall thickness (Figure 5.24). The stability of the whole structure will then be improved. The wedging must be carried out by placing the fill stones manually one by one. In the case of a coursed masonry, the larger stones will be used, while avoiding having too much height difference between them, to ensure a 'platform' that is more or less levelled. In the case of an incertum opus, the largest blocks will be assembled with side contacts as tight as possible.

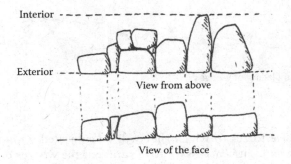

Interior

Exterior

View from above

View of the face

Figure 5.24 Setting of the foundation stones.

5.2.4.1.1 Foundations on soft soil

It is also possible to dampen the soil and hit foundation stones that have been laid flat with a rammer: this operation will increase the compaction of the soil. Alternatively, a light compaction plant may be used. However, it must be understood that these measures will only compact a shallow depth of soil, much less than the depth of soil that will experience large increases in stress when the wall is constructed.

When the soil is too soft, the stones can be placed vertically (Figure 5.25a), pressed down in the earth or pressed together as a block using wedges; this is described as a 'raft foundation'. There are also arrangements made of large stones acting as micropiles, connected by stones laid

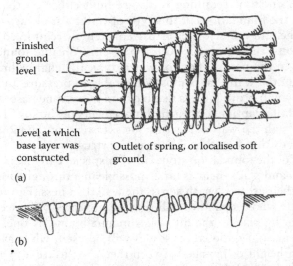

Finished
ground
level

Level at which
base layer was
constructed

Outlet of spring, or localised soft
ground

(a)

(b)

Figure 5.25 Modification of the foundation construction on soft soil. (a) Vertically placed stones. (b) With stones acting as micropiles.

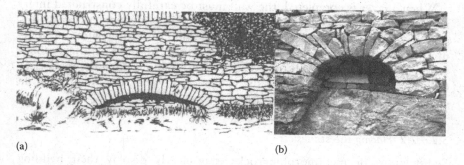

(a)   (b)

Figure 5.26 Arches spanning a weak zone. (a) Drawing. (b) Example in limestone (Hauterives, France).

on edge in an arc (Figure 5.25b). Finally, highly localised weak zones may be partially spanned by the construction of relieving arches (Figure 5.26).

5.2.4.2 *The wall itself*

5.2.4.2.1 *Beds of stones*

The wall is built layer by layer, in successive courses. The courses are perpendicular to the batter (Figure 5.20), at least in the first third of the wall. It is logical to place the stones in sequence along the length of the wall, for it may be difficult to find stones that will fit exactly a gap between two already placed. The external facing and internal facing can be built simultaneously by two builders working on each side, together, to promote better placement of stones to connect the two faces inside the wall. The stones within a bed should be adjusted so that they are as far as possible in contact with each other, and locked together by their shapes and by friction between them – it should not be possible to withdraw any stone from the completed face of the wall. The lateral contact ensures the friction between the stones and contributes not only to the cohesion of the whole structure, but also to the aesthetics of the facing. The vertical joints to the rear should be filled with locking wedge stones (Figure 5.27).

In a coursed masonry wall, the horizontality of the beds along the length of the wall must be ensured, not only out of a concern for aesthetics, but to prevent sliding along the length of the wall which could result in collapse.

Figure 5.27 Wedges hold the back ends of the stones in place – view from above.

When a layer is completed, the wall must be carefully constructed internally before the next bed is laid. The internal organisation of the wall is critical to stability. Generally, one aims to minimise the volume of voids in a wall, but it is also important that drainage is allowed, so these stones will typically be typically be of a size at least 40 mm. Filling the interstices ensures better friction between the stones of the wall, thus increasing its cohesion and density.

5.2.4.2.2 Placing the stones

Sedimentary or metamorphic rocks that clearly display their bedding planes (e.g., schists, limestones and some sandstones), and are hence usually built as horizontally coursed masonry, should be placed 'flat', that is, on their natural base, in the direction of their stratification (Figure 5.28), so that they are taking load within the wall in the same direction as they were when they were formed in the ground. This gives the best resistance to compression. If the builder chose to place a stone vertically, then there is a risk that the stones will be split by the pressure acting on them in an unfavourable orientation. The stones should be oriented so that the bedding planes are perpendicular to the wall face, so that the adjacent stones give some pressure to prevent delamination of the stones (Figure 5.28). It is important that such stones are not too high, nor create running joints. If the entire construction is made with vertically orientated stones, then concerns related to the function of the wall or the nature of the stone have been regarded as more important than this normal consideration.

5.2.4.2.3 Rule of crossing joints

It is necessary to avoid the alignment of vertical joints (running joints). To do this, the stones will be built so that their joints will be staggered or

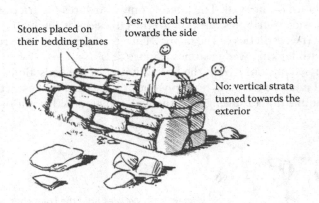

Stones placed on their bedding planes

Yes: vertical strata turned towards the side

No: vertical strata turned towards the exterior

Figure 5.28 Stones placed horizontally and vertically.

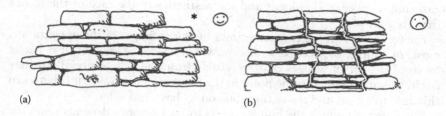

Figure 5.29 The rule of crossing joints – giving priority to making the joints out of line (a), so that running joints (b) are avoided.

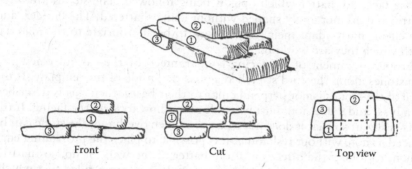

Front   Cut   Top view

Figure 5.30 Crossing joints both horizontally and vertically.

crossed (Figures 5.29 and 5.30). It is not always easy to avoid the appearance of running joints – standing back and observing the work during the building allows monitoring of this, and so avoiding the error.

The courses often consist of variable height stones, and two or even three thinner stones may be placed adjacent to a thick stone, so that the next bed then passes from the top of the thick stone onto the topmost thin stone. This requires the thin stone to be precisely the correct thickness, so that a stone can rest on it and on the thick stone, to prevent a running joint; this may require resizing of an appropriate stone. Alternatively, a stone may be found that fits the step, as indicated by the asterisk (*) in Figure 5.29.

Just as it is important to avoid vertical running joints on the visible face of the wall, so it is equally important to avoid running joints within the thickness of the wall, both horizontally and vertically. The goal is to achieve a cohesive structure, and this requires the avoidance of running joints even within the hidden part of the structure (Figure 5.30).

5.2.4.2.4 Facing stones

The exterior facing and interior of the wall should be built with the same rigour, although the appearance of the facing will receive more care in ensuring its proper alignment and the aesthetics of the face of the stones that are part of it.

Facing stones, which can sometimes be cut to improve the appearance, must, of course, be wedged with care, when needed. However, it is better to avoid the use of wedges that are visible from the face, especially if they could be withdrawn at some point in the future (Figure 5.31): it is better in this case to adjust and resize the stone on its base and sides.

The extent to which the builder needs to resize stones depends upon the wall type and the stones available. Generally, this can be limited to a little cutting, to adapt the stone to the shape of the space it is to fit, or to eliminate the defects of the face. However, it may be required to have very thin joints at the face, no matter which opus is being followed. The stones are often brittle and are not easily shaped without being shattered. The builder, and the client, must adapt their expectations to be appropriate to the material with which they are working.

Proper alignment of the facing is guaranteed by the careful placing of the stones along the cord stretched across the inside of the template. If the face of the stones is not perpendicular to their bases, then this is described as a bias, and a decision must be made about how to form the facing. If the bias is slight, so that it does not cause too much overhang, the stone can be placed as it is, without reshaping. It is possible to place the protruding edge down, to follow the direction of the batter of the wall, or up, so that the top of each course is aligned with the cord. It is recommended that which-ever option is selected, it is followed consistently throughout the face of the structure to ensure a harmonious appearance (Figure 5.32).

When their face allows, the stones should be laid as partial or full through-stones (Figure 5.33), that is, with their greater length in the thickness of the wall. This arrangement allows each stone to provide a greater resistance to the distortions arising from the earth pressure on the back of the wall and achieves a maximum cohesion between the facing and the interior of the wall. This rule is often called 'through-stone implementation' but does not replace the through-stone itself (see Section 5.2.4.2.4). If the quality of a

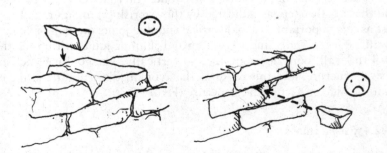

Figure 5.31 Visible wedges.

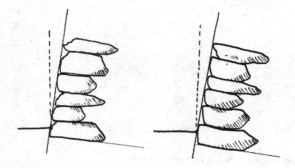

Figure 5.32 Orientation of the bias of the facing stones.

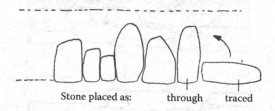

Stone placed as: through   traced

Figure 5.33 Placing stones as partial throughs, or traced.

facing leads to stone being placed as a traced stone, that is, with its longer side in front, to avoid a running bond within the wall, it should be tied in with a through or partial through above it on the next course.

5.2.4.2.5 Through-stone

Use of regularly placed through-stones ensures good cohesion and stability for the wall. The relative number of through-stones depends on stones that are available, but the more they are, the more the structure will present internal cohesion and thus resistance to backfill thrust. If there are not enough stones of sufficient length, this may be remedied by arranging two or more stones as complementary through-stones, one beside or above the other (Figure 5.34). This is called *en epingle* in France, and as partial or three-quarter throughs elsewhere. The stones must fit snuggly alongside each other, and be held together by stones placed above and below, so that they prevent the inner and outer facing from separating, just as a full through-stone would do.

When the wall is very thick, it is necessary to create a chain of stones in through to connect the two facings (Figure 5.35) – this is almost inevitable in a very large wall. They should be placed in the same course and crossed by the top course. These stones will also have as large a contact area as

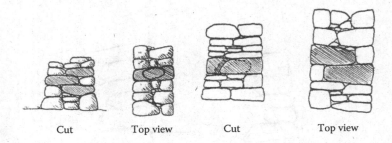

Cut   Top view   Cut   Top view

Figure 5.34 Overlapped three-quarter throughs.

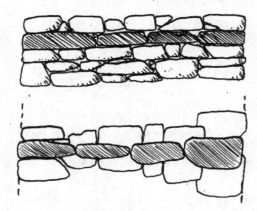

Figure 5.35 A chain of overlapping stones linking the front and back faces.

possible, to increase the friction between them. A staggered arrangement of stones may also be used.

5.2.4.2.6 Internal organisation of the heart wall

The stones of the outer facing are placed first, followed by those of the rear facing. Then the inner part of the wall has to be built without changing the wedging of the stones already placed. The internal organisation ensures the cohesion of the whole. It is completely unacceptable to fill the wall by just tipping stone in between the facings from a bucket or wheelbarrow. It is very important that each stone is arranged and wedged methodically one by one, with the same rigour applied to building the facing. For this reason, and to ensure good drainage, the stones used should not be too small.

First, large stones should be chosen to properly cross the waiting bed between two facings, mixing different sized units as necessary; then the voids are filled with small stones that fill the remaining spaces. If rounded pebbles are used, particular care must be taken to ensure that they cannot 'roll like marbles', destabilising the entire structure. Well packed fill will take part of the weight of the wall above it, and should be of good quality stone, packed so that it will not become rearranged as the wall is built. It is much better to have stone that is rough and angular rather than smooth and rounded.

Filling the stones between the two facings should never go above the upper surface of the facing stones; otherwise it will hinder the installation of the next course.

5.2.4.3 The coping

The coping concludes the construction of the wall: its roles are to link the two facings, level the upper part of the wall and protect from damage the lighter stones that are normally used towards the top of the wall.

The coping stones are placed along the cord to the final height of the wall by adjusting the nearest, and if the style of coping flat stones, they must be wedged very carefully, possibly using thinner stones to make up for differences in thickness of the coping stones, so that the final surface is level. The few remaining gaps between them must be filled, but large numbers of small stones should be avoided, as they will be exposed and could be easily removed.

In the case of a flat coping, the weight of the stones should be sufficient to make them difficult to move (Figures 5.36 and 5.37). They should be based on a stable course and bind together both faces of the wall. Even if

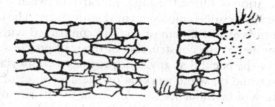

Figure 5.36 Covering with large flat-surfaced stones, front face and cross-section.

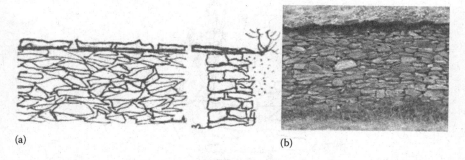

(a) (b)

Figure 5.37 Covering with thick slab stones. (a) Drawing front and cut. (b) Example in schist at Faux, France.

Figure 5.38 Covering with vertical stone, front face and cross-section.

there are not enough long stones available, it is important that the coping is made of thick and heavy elements, arranged so that the stones underneath are overlapped and hence tied together.

In the case of a vertical covering (Figure 5.38), the stones must have more than one contact point so that they serve their purpose of tying together the front and back of the wall. According to the operating mode selected, the stones can be arranged either as vertically as possible, or slightly inclined, and with or without level tops. Vertical wedging may also be provided by adding stones carved to suit their purpose and pushed down in the vertical joints.

5.2.4.4 The backfilling

This is the operation of filling the gap with earth between the inner facing of the wall and the existing embankment (Figure 5.39). The earth is generally deposited in layers of 20 cm or so, and compacted with a manual or mechanical rammer as the construction of the wall progresses, so that the upper surface of the backfill at any time provides a comfortable height for the builders to stand on while working on the next stage of the wall; this working height will be influenced to an extent by the size of stones used, but waist height is more normal than the knee height shown in Figure 5.39.

Figure 5.39 Backfilling.

Good compaction reduces the risk of collapse of the embankment during rains, regulates the water retention and the migration of fines within the wall, and prevents undue settlement of the retained ground surface. It also maximises the shear strength of the backfill.

If possible, waste stone from shaping of the stones used and gravel-sized fragments (not to be confused with the wedging stones; see Section 5.2.1.3) should be put into the base of the backfilling and behind the back of the wall. This is the drain that will recycle waste stone while increasing the drainage capacity of the structure. It will filter the soil particles and keep them from clogging the wall. It is strongly recommended when backfill is a clayey soil. When available stones for backfilling are layered, they should be placed horizontally, because if placed carelessly they can add an extra thrust on the inner side of the wall.

5.2.4.5 End of the work

After the construction of the wall, the ground surface above and in front of the wall must be levelled as required; the working and storage areas must be cleared and cleaned, and waste and surplus material removed from the site.

5.3 SUMMARY

This chapter has set out the rules that must be followed to obtain a strong and ductile drystone retaining wall, in a safe and efficient manner.

ACKNOWLEDGEMENTS

This chapter is based on rules of professional practice developed from Chapter 3 of the book *Guide de bonnes pratiques de construction de murs de soutènement en pierre sèche* (A Guide to Practice in the Construction of Drystone Retaining Walls), written by a consortium of French organisations under the guidance of Jean-Claude Morel and published by École Nationale des Travaux Publics de l'État (ENTPE), part of the University of Lyon (CAPEB et al. 2008). Chapter 3 was written by Artisans Bâtisseurs en Pierres Sèches (ABPS – French national association of drystone wallers who have been working closely with ENTPE since 2002). The authors are most grateful for their permission to draw on this text and to reproduce some of the illustrations. The drawings in the illustrations are by Zarma, courtesy of ABPS, who have also kindly provided many of the photographs in this chapter.

They would also like to thank William Noble, Senior Examiner for the Dry Stone Walling Association of Great Britain, for his help in reviewing this chapter.

Chapter 6

Assessment

Drystone retaining walls may be hundreds of years old, and many changes may have taken place since they were built. The most obvious change in loading is due to wheel loading from modern vehicles, which can be particularly problematic if a road passes close to the top of a wall; however, the loads imposed by modern agricultural machinery are also very high, so few walls are immune to the increases in imposed loading.

Parapet walls are a common feature in some areas, yet they may be removed if their condition deteriorates – which could reduce the stability of a retaining wall because it reduces its weight. Often vegetation is allowed to grow close to the top of a wall, which as well as adding a vertical load, can lead to an unfavourable lateral load when the wind is in the wrong direction. A wall might have had another wall built on top of it, to allow the retained ground to be levelled out. The difference in appearance of the later addition, which reveals this has been done, also implies that the original wall has not been widened to support the additional height of retained fill.

Perhaps the most serious change in loading may come from a build-up of pore water pressure. Even if the fill behind the wall started off as a free-draining granular material, many years of soil and dust being washed through it, together with weathering and precipitation of minerals, could result in the permeability being dramatically reduced, so that pore pressures build up until failure takes place. The most common cause of increased pore pressures, however, is the pointing of drystone walls, which can remove their free-draining nature if steps are not taken to ensure good drainage.

Drystone walls also change over time as the stone they are built from deteriorates. Even if the stone may have been in the ground for millions of years before it was built into the wall, once it is exposed to sun, wind and rain it may undergo chemical and mechanical weathering. Chemical weathering changes the nature of the constituent minerals, perhaps through reactions with water, oxygen or weakly acidic rain, to a material that may not have the same strength as the original mineral, and may have a greater volume so that it breaks up the stone as it expands. The most common type of mechanical weathering is freeze–thaw – the expansion of water in pore space or cracks within the stones can break the stone into pieces. Every cycle of freezing and thawing may weaken a stone further. A dense stone will resist water penetration, as well as being stronger, and so will resist freeze–thaw weathering.

Investigation of aspects that could change with time is important, because some changes could lead to a wall becoming unsafe if they continue. Determining whether or not they will do so also depends on an assessment of the margin of safety with which the wall is functioning at present. Chapters 2 through 4 indicate the calculations required, but the data needed for those calculations are not readily available when assessing an existing wall. The most fundamental requirement is to determine the geometry and the form of the construction. This is difficult because only the outer face of a wall is visible. Endoscopes are used to investigate the voids in drystone walls for signs of living inhabitants, such as bats, but may also be used to see some of their internal construction. Close visual inspection, with the aid of a powerful torch, can reveal something useful, and a stiff wire used as a probe can be helpful too. However, the most powerful nondamaging technique to date is thermal imaging; radar and sonic investigation is confused by multiple reflections, because a drystone wall contains so many surfaces. Beyond this, the information the engineer would wish for cannot be obtained without drilling behind the wall, and possibly dismantling it.

6.1 THE SCOPE AND PURPOSES OF ASSESSMENT

6.1.1 Definition of function

An appropriate assessment cannot be made unless the function of the wall has been correctly identified. In many cases a wall is providing more than one function, and missing one may have serious consequences. Here are some functions to consider – but it is important to be open to the possibility of others.

1. To provide an area *in front of* the wall which is level or gently sloping
   This is in fact a very broad category of function, but is one of the two fundamental functions of a retaining wall. There are many purposes to which this area may be put. If the ground is cultivated, then there may be people there occasionally, whose safety would be of concern; there may be agricultural machinery causing damaging vibrations, or possibly impacting the wall; the ground immediately in front of the wall may be ploughed, risking a reduction in pressure in front of the wall which could lead to bearing failure at the toe. If this area is for a roadway, then traffic may interfere with inspections, and a temporary closure may be needed just for the assessment; there could be damage due to vibration or impact; there is a possibility that services (pipes, cables or drains) have been or will be installed in

trenches cut in front of the wall; highways authorities may require a level of assurance of safety that requires calculations which depend upon detailed data which could be difficult and expensive to obtain. If the area is for a building or a garden, then access may require permissions, or be very limited if the building is close to the wall. In the United Kingdom a retaining wall is usually owned by the owner of the land it is retaining, rather than the land in front of it, yet a collapse may have more serious consequences for those in front of the wall than for those above it. Two retaining walls at different locations near Bath in the United Kingdom failed in the summer of 2014, causing very serious disruption to traffic for many weeks, a far greater consequence than the direct effects on the owners of the structures. It may therefore be important to understand the relationship between the ground in front of the wall and that retained by the wall, and the respective ownerships that may be different from those when the wall was originally built.

2. To provide an area *behind* the wall that is level or gently sloping

This is the second fundamental function – both tend to imply that without the wall there was a natural slope, but in this case the ground behind the wall may be entirely made up, with either another wall or a gentle slope forming the other side of the platform supported by the retaining wall. As with the first category, the purpose of the level ground must be established. Even the simplest case of agriculture brings with it the possibility of heavy agricultural machinery close to the top of the wall, causing local increases in pressure on the back of the wall that could lead to collapse; tractors get larger as each decade passes, so the absence of problems in the past is no guarantee of an absence of problems in the future, particularly given the possible deterioration of the wall. Impact loads may also be a problem, if there is a parapet or fence at the top of the wall. If there is a road on the ground behind the wall, it makes a big difference whether or not there is a verge or barrier preventing vehicles from getting too close to the top. However, even if there is a barrier, an out-of control vehicle could break through it resulting in large concentrated loads close to the top of the wall. Many roads have services such as water, gas or electricity supplies or drainage, in buried pipes or conduits, which may be in the fill behind the wall rather than further away from it. This could cause considerable problems for working on the wall, and if one of those services is a leaking water pipe, it could lead to soil being washed through the wall. If the wall retains a canal, then the loading on the wall will be constant because any barge on the canal displaces its own weight in water, but if the canal should be leaking there could again be erosion problems. Waves could cause temporary higher loads, or overtopping of the canal resulting in soil erosion. However, the most likely cause of leaking in such a situation would be movement of an inadequate wall. If there are buildings on the ground at the top of a wall, then the sensitivity to movement may be high. Sometimes one wall of a building rests on the top of the wall. In this case the load from the building adds to the self-weight of the wall, and may actually be increasing its stability against sliding or overturning failure – which could subsequently occur were the building to be removed, or at least the margin of safety could be considerably reduced, resulting in some movement.

3. To provide a dead-weight at the foot of a slope

A deep-seated slope failure can often be stabilised by placing weight at the toe of the slope. In this case the wall would be holding in place a significant weight of fill. If a decision was made to remove the wall and replace or rebuild it, then the removal of material could result in a failure of the slope.

4. To provide an aesthetically pleasing part of a landscape

This may be an agricultural landscape, whether arable, pastoral or horticultural, or it may be part of a garden, whether large or small. In either case, an alteration in the aesthetics of the wall could have a serious effect on the landscape. It is possible that replacing a dilapidated wall with a neat new wall could ruin the character of the landscape, and considerable skill may be needed to produce an acceptable repair or reconstruction. The ha-ha is a stone wall lining one side of a ditch, to control livestock where a visible fence or wall would be intrusive; in this case, the prime virtue of the wall is that it is not seen.

6.1.2 Definition of need

Once the function of the wall has been identified, it is essential to be clear why the assessment is being made. Here are some examples – all centre on function and safety.

1. *Will property or services be damaged if the wall deforms too much?* Some walls will serve a function that would result in a failure even without the wall collapsing, for example, if buildings or services are based on or near the retained fill. This would be strictly classified as a 'serviceability' failure, for which a lower margin of safety is normally required. The presumption is that deformation beyond the normal limits is less serious than a complete collapse – but too much movement could fracture pipes or lead to structural damage nearby. The wall would still have to be rebuilt, and so would to all intents and purposes have 'failed'. In such cases the need for assessment is proportional to the importance of what the wall is supporting. In general, a safe margin against collapse would also ensure that deformations would only be modest, so the assessments required may seem to be similar; but it is also true that walls of poor construction can

undergo significant deformations without being in danger of collapse, especially if lower quality stone and construction techniques are hidden behind a carefully built facing.

2. *Is the wall safe for people to be in front of or at the top of?* The likelihood of someone being in front of a wall or on the ground it retains when it collapses depends on its location and purpose, but if there really is no chance at all of someone getting hurt by a wall collapsing, and there are no issues covered by the previous paragraph, then it is unlikely that anyone will want to have an assessment made. If the wall may present a hazard, then given that the existing condition is that the wall is standing, the assessment needs to consider what changes may be happening or about to happen that could change this state. At the design stage, we aim for a margin of safety that allows for the imprecision of design parameters, especially for material properties. Once a wall is built and loaded, then many of these uncertainties have been fixed, for better or worse. It may be assumed that because the wall is standing it is safe, but it may have not yet experienced a full design loading. If it has, then we might say that it has been proved to function, and the issue of margin of safety is no longer relevant. However, one always expects there to be some spare capacity to allow for deterioration, or for overloading. Assessing that spare capacity will usually require a detailed investigation. However, the ductility of drystone walls means that the wall will undergo significant deformations before it collapses, so that an assessment of risk of collapse can be carried out by an examination for excessive deformation, backed up by assessments of quality of construction and condition of materials.

3. *Is the wall at risk of collapse due to progressive deterioration?* In this case we are concerned not just with safety, but also with the nuisance value of having the area at the top of the wall or in front of the wall unavailable until a repair or replacement has been completed. A single collapse could close an entire road for weeks, and there may be hundreds of metres of wall large enough to pose this risk on a single route. Establishing an efficient and effective protocol for inspection and assessment can therefore be very important. The difficulty here is that visible deterioration at the face may not be as significant as hidden deterioration in the body of the wall, and especially deterioration of the smaller fill stones; these small stones lose a significant proportion of their volume to weathering more quickly than the larger stones at the faces, even if they are of the same quality.

4. *Is the wall at risk of collapse due to changes in loading?* This type of assessment can be the most difficult, because there is no reassurance from the fact that the wall has been doing its job so far. A change in load may be due to general changes in wheel loadings; this may be a change in permitted loading on an individual tyre, or a change in

the combinations of loadings from a closely spaced group of wheels. For example, to allow heavier vehicles, an additional axle might be permitted so that the load is carried on three axles rather than two, so that the wheel load is lighter and overall damage to the road pavement is reduced. The effect on structures in general may be insignificant, but the effect on a drystone retaining wall could be to load a greater length of the structure, so there is less potential for the structure to redistribute the load through the interactions between the stones. Localised faults that did not matter before could then lead to collapse. There might also be a need to make a special assessment of a wall because an abnormally heavy load is to be carried on a road it supports, or if a tree that is growing close to the top of the wall could transfer significant horizontal loading during strong winds. It may be impossible to give assurance that a wall will be able to sustain increased loading without knowing the engineering properties of the backfill, the geometry and construction and the properties of the stone with which it has been made. To achieve this may require the dismantling and subsequent reconstruction of part of a wall. For a short section within a long length of wall, the cost of such an investigation may be justified, especially as the drystone construction may be rebuilt to restore the original appearance, possibly with some new stone being incorporated.

5. *What might be done as a consequence of the assessment?* Before an assessment can be designed, it is necessary to have a clear idea of what might be done as a consequence of the assessment; otherwise time and effort may be spent on investigations with no clear purpose. This is part of the definition of the need for the assessment. The possible actions might be briefly summarised. The first option, and usually the most desirable, is to do nothing. Then there is a possibility that some local repair may be needed, though this is most likely to turn into local dismantling and reconstruction. Grouting and anchoring have been used, but both can have serious problems.

6.1.3 Definition of information required

The specific information required will depend on the function of the wall and the purposes of the assessment. However, even the most basic assessment will require determination of the geometry of the wall – ideally beginning with an elevation, showing the wall from the front with its length measured, and its height shown at different points along its length. Of course, if there is just a short length of wall with a uniform height then the information becomes simpler, but the investigation should then record what is at either end of the length being inspected because this may be an important factor.

From here on the investigation rapidly becomes more complicated. The next step is to determine the profile of the face of the wall. Because the wall face is made of lots of individual stones, each of which might have a very uneven face, it is necessary to be clear about the level of detail required. Laser scanning might provide a very detailed picture, and if the scan interval is sufficiently fine, it would be possible to identify individual stones from the resulting scan. This might be useful for comparison at a future date to identify any major deformations that have taken place in the interim, but it does not provide any useful information for an assessment, except for the fact that coordinates of identified points may be taken by interrogation of the scanned data. The more important matter of identifying which points should be measured remains. If the face is approximately flat, it will only be necessary to determine its height and the angle it leans from the vertical. If the face is not planar, whether through construction or as a result of deformation, then it is likely that several vertical profiles will be needed to define its shape sufficiently accurately to allow worthwhile analysis.

Any analysis also requires the position of the back of the wall to be determined. This is much more difficult, as the back of the wall may not be accessible. Even if a small excavation can be made safely at the top of the wall to determine the position of its back face, there is a strong possibility that all that is revealed is a parapet wall that has had some fill behind it, and the true back of the wider retaining structure remains hidden at greater depth.

While considering the back of the wall, there is also a need to determine the properties and condition of the retained fill. In many drystone retaining walls, the fill immediately behind the structure is made up of broken stones and stone fragments that were not used in the wall. The fill near the surface may be topsoil, or layer upon layer of road construction, so it is not going to be feasible to ascertain the geotechnical engineering properties of the backfill without a proper soil investigation; this is likely to include drilling and sampling, followed by laboratory testing. Given that the position of the back of the wall, and hence its width, also requires invasive investigation, it is likely that the first stages of any investigation will focus on what can be ascertained from the face of the structure, and will focus on evidence of construction style and quality, and on evidence of changes that have taken place since construction.

6.2 METHODS OF ASSESSMENT – OBSERVATION AND INTERPRETATION

6.2.1 The stone itself

The primary concern with the stone is that it continues to have high compressive strength and stiffness. Most stones undergo weathering when exposed to wind, rain, abrasion by wind-blown particles and freezing and thawing. Vegetation can also contribute to the deterioration of stone, principally through the physical action of roots. The way in which the exposed stone deteriorates over time may not match what is happening to the stone hidden behind the wall face, which is doing most of the work. Wind and rain, as well as providing mechanical weathering of a kind that will not be experienced by the hidden stone, tends to remove any weakened material, and it may not be obvious that this is happening. Wetting and drying can lead to the progressive deterioration of the strength of some of the types of stone that might be used for drystone retaining walls.

The stone that is exposed at the face may be doing little more for the stability of the structure than adding weight – the actual points of contact between the stone may be some distance back from the face and exposed to the air only through a relatively thin gap. Provided that there is still sufficient weight to resist sliding, the loss of stone at the face that is not contributing to the transmission of vertical load may have very little effect on the resistance to overturning, and might even improve the distribution of bearing pressure on the foundation by allowing the centre of mass of the wall to move backwards.

In limestone walls, precipitation of calcite has been observed where the stones are in contact with each other, so that the long-term effect is a strengthening of the connection between the stones and the formation of a hard layer that protects the parts of the stone that are transmitting the forces. Considerable care is therefore needed in observing and interpreting the deterioration of the stone.

There may be stones that are cracked. Close inspection of the cracks might reveal signs of when the crack took place, if the fresh stone surface on the crack changes colour or weathers with time, or the crack fills with dust or dirt. Occasional cracks may occur during the construction of the wall, as increasing load results in a bending moment in stones that are not adequately supported (Figure 6.1). Normally this is not a problem if the

Figure 6.1 A number of cracks are visible in the stones in this photograph, all of which probably took place as the wall was being built.

crack is visible at the face. Even if such a crack results in a running joint, it will only be very short if the rest of the wall has been well constructed. A cracked through-stone would have a greater consequence, resulting in loss of connection between the front and the back of the wall, but such a crack would be within the body of the wall, and not visible at the face. A fresh crack in an old wall could be a matter of concern, as it could be a consequence of stone deterioration, external loading or settlement or subsidence causing a redistribution of load within the wall. A series of cracked stones is likely to indicate a serious problem of overloading or foundation movement.

6.2.2 Geometry

In most cases, the first measurements will be taken with a long tape to obtain length and height of the structure, but a reflectorless total station will allow much more detailed information to be obtained, probably more safely. This instrument is an electronic theodolite that incorporates laser distance measurement as well as the recording of angles, so that it can store calculated coordinates of the points being observed. Such instruments require a specially designed reflector to measure to, often attached to the top of a pole that is placed on the point being surveyed. Reflectorless instruments dispense with the need for this, using a powerful visible light laser that can measure a distance to most surfaces. This means that a wall may be surveyed from a safe vantage point, without the need to make physical contact with the wall, or even approach it closely; this could be particularly advantageous if the wall is next to a road.

If a reflectorless total station is not available, or there is no one available with the expertise to use a hired instrument, then tape measures and spirit levels are likely to be used.

If the face of the wall is not planar, for example, if the face angle varies over its height and along its length, or if there is bulging, then recording requires much more detail. The reflectorless total station will still work well, but it may be difficult to tell if there are distortions from a distance. If it is not possible to approach the wall closely enough, then it is best to observe horizontal lines of points at different levels, and vertical lines of points at different sections along the wall, so that distortions might be identified once the data have been processed. Such interpretation requires an understanding of the accuracy to which the wall is likely to have been constructed. Indications of this may come from the overall apparent quality of the construction and the extent to which the visible faces have been worked to give the wall a flat face.

Working with tape measures is easiest with clear access to the top of the wall as well as to the ground in front of it. Then a plumb-bob (a line with a weight on the end) can be hung from a piece of timber projecting from the top of the wall, with the weight set to the toe, or a distance in front of the toe if

Figure 6.2 A levelling staff used to assist observation of a bulging section of wall adjacent to a wall that has failed.

a bulge must be accommodated. Then horizontal offsets from the line to the wall face may be measured at measured heights to determine the face profile. A vertically placed levelling staff (Figure 6.2) held to the projecting timber would assist this process, and for higher structures a ladder may be needed.

An alternative approach is to use a laser scanning system. In some ways the actual operation of the system is easier than using a total station, but the equipment is expensive and the software processing is time consuming. The instrument can be set to scan the wall horizontally and vertically, to produce a 'point cloud' of the coordinates of the surfaces that have reflected the laser. Any object that gets in the way will be measured instead, but careful choice of instrument location and observation time (to avoid traffic and pedestrians) would avoid this problem. The accuracy of such points should be to within a few millimetres, which is more than adequate. The frequency of measurements can be set, so that the points are not unreasonably close together; however, because they are simply on a grid, rather than choosing suitable points on the face of each stone, a high density of points is likely to be needed. Then points to be used in defining the geometry are extracted from the point cloud back in the office. The system has the advantage of the reflectorless total station in measuring from a safe distance, and will probably be quicker in the field, but will give much more information than is really needed for defining the geometry. However, if the point cloud is sufficiently dense there will be no need to measure the dimensions of individual stones for surveying the construction, as both overall and detailed geometry will be obtained in one scan.

6.2.3 Details of the construction

Although a laser scanner can be used to obtain some details of the face, as described previously, a series of good quality photographs taken with the camera pointing directly at the face will provide adequate initial information. A levelling staff resting against the wall within the image is a convenient way to provide a scale. The size, orientation and variety of the stones should also be noted when still in the field, rather than relying on photographs, as more information will be gathered with the wall in front of you than working from a photograph. Chapter 5 may be used as a guide to the characterisation of visible aspects of the construction, and to what is being looked for. Any coursing of the stones or vertical running joints will also be described, and the spacing between such features should be noted. Particular attention should be given to evidence for different phases of construction or repair – for example, stone of different ages or changes in the tightness or pattern of the construction. Figure 6.3 shows a very clear example of a poor connection between poorly constructed walls.

If the face of the wall has been constructed to form a flat plane, then any stones that are not flush with the face will be obvious. It is necessary to look for any clues that might indicate whether such stones have moved, or whether they are perhaps through-stones that have been deliberately left projecting from the face. If the face of the wall is not flat, it is necessary to explore why this might be the case. Sometimes this is simply due to construction without due regard to this aspect – walls built by amateurs are

Figure 6.3 A very poor junction between adjacent sections of a limestone wall.

most likely to have uneven faces. Often the cause will be movement during or shortly after construction, and it may be clear that this comes from the sliding of part of a course of stones, or bulging either locally or along an entire length of wall.

Gaps between stones should be probed with a torch, and possibly a stiff wire, to determine if there is any filling in the joints and the sizes of both the main stones and of any filling stones. Occasionally, there may be places where the entire width of wall is apparent if the fill stones are large. An endoscope may also be used and is a routine tool for examining walls as wildlife habitats. Such close examination will also reveal if a wall has been assembled using mortar, while trying to maintain the outward appearance of drystone. This is becoming a common practice where the materials are relatively cheap and the skilled builders who are able to work quickly and efficiently in drystone are not available, yet the surroundings call for drystone. Ignorance of the rules of drystone retaining wall construction often result in a very poor pastiche of a real drystone retaining wall, so the appearance may raise suspicions even before any detailed inspection is carried out. Such structures may have adequate strength, but do not have the ductility of drystone, and may not even have adequate drainage, leading to the build-up of pore water pressures and ultimately failure.

Thermal imaging can reveal aspects of a wall's construction that could not be discovered otherwise without dismantling. Temperature variations within the ground are much less than the variation in air temperature, as the ground acts as an insulator, and because of its heat capacity it can change temperature only slowly. At a metre depth, temperature changes during the course of a day are very unlikely to have any effect. Drystone retaining walls have earth resting against them, and so are connected with material that is at a more stable temperature than the face of the wall, which is exposed to the air, rain, and if it is facing in the right direction, to the heat of the sun. The temperature of stone in good contact with the retained soil will therefore be more stable than that of the surrounding stone which is not. A through-stone that extends from the retained fill right to the face will be the most stable. When the air temperature is lower than the ground temperature, the face of such stones will be warmer than the face of surrounding stones and vice versa. Investigations have shown that this effect is clearest in the morning after a cold night. Later in the day, once the air has warmed up, especially if the sun is shining on the face of the stones, the stone with better contact with the soil may be cooler than the surrounding stone, but the thermal conductivity of individual stones plays a larger role. The surface of the stones may heat up to similar temperatures in the sun, and the effect of the contact with the ground becomes secondary. As thermal imaging cameras may have a temperature resolution of 0.1°C, subtle differences can be detected. It should be noted that as we are only interested in temperature differences between objects in the same view, the absolute accuracy of temperature measurement does not matter.

The technique can give clear results for through-stones, but good bonding between the front and back of the wall will also be apparent. Where the wall is made of two faces and in-filled, then the thermal conductivity is reduced, so the through-stones that are essential to this form of construction stand out particularly clearly. Variations in the infill also become apparent, the most obvious being where the front face is actually beginning to separate from the back face and infill, which is the first stage of the development of an unstable bulge. Images of sections of wall that include repairs have shown good continuity in the new repair, but poor continuity in the old wall sections left on either side.

Figure 6.4 shows a pair of images of a granite retaining wall, taken in the afternoon when most of the stone has warmed up. A number of stones are evidently cooler than the surrounding stones, suggesting that they are

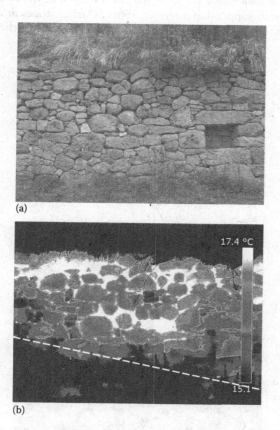

Figure 6.4 (a) Natural light and (b) thermal images of a wall in Lozère, France. In the thermal image, the lightest tones indicate the highest temperatures. Note in particular the pair of darker stones just above the centre of the thermal image, which are cooler.

through-stones. The stones in Figure 6.4b are colder, below a line that slopes down towards the right (a dashed line has been superimposed on the photograph); this line is apparent in the construction in the natural light view. This suggests the possibility that the lower courses may be constructed close to a rock outcrop, while above that level is a full thickness of drystone wall. The wall is shown during construction in Figure 5.14, and the substantial rock outcrops immediately behind the wall can be clearly seen – not all of this structure is retaining significant earth pressure. It might also be possible that the cool rocks have simply been in the shade all day so far, as this photograph was taken in the afternoon. However, the wall is south facing, and this was not the case.

Figure 6.5 shows a pair of images of a wall made of schist. A decorative band of vertically orientated stones probably marks a transition from a revetment built in front of a rock face (to form part of a terrace at the foot of the wall) to a retaining wall supporting a terrace above. Darker stones at intervals on a level approximately two thirds of the height may be a line of through-stones.

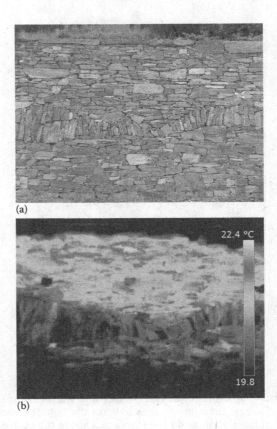

(a)

(b)

Figure 6.5 (a) Natural light and (b) thermal images of a wall in Lozère, France.

An old technique for gauging the soundness of a wall is to hit it with a wooden mallet, or something similar. Stones that are not held in place will become immediately apparent. A solidly built wall will 'ring', whereas a wall that has many soft points of contact will absorb the sound. If the front of the wall and the back of the wall are not well connected, a hollow sound may be apparent. Of course, this would not be attempted if the visual appearance of the wall suggested there was a risk of it collapsing. The investigator should start by testing large stones in sections of wall that appear to be of good quality to learn what sound to expect.

Some investigations have been carried out using ground penetrating radar (GPR), but the very large number of air-stone surfaces has led to a very confusing picture. It is possible that further progress might be made using this technique, or using seismic investigation methods, but at present the thermal investigations give the most useful information.

It must be accepted that there will be situations that require detailed investigation, which might require partial dismantling and a full investigation of the backfill materials. This might be the case, for example, if new loads are to be applied to a structure, or developments in its vicinity increase the consequences of failure. If this is to be done, it would usually be advisable to provide temporary support to the face either side of the structure being dismantled; this is because removing part of the wall is likely to transfer load to adjacent sections. Investigation would then usually begin with partial excavation of the backfill, ensuring a safe slope angle or good support for that which is left in place. Such investigation must be done by those with appropriate specialist experience of ground investigations, and the temporary support structure must be of sufficient strength, and sufficiently well-supported itself, so that it is capable of carrying the large loads that might be placed on it as the soil arches over the section being examined.

6.2.4 Defects

Having completed observations, it is necessary to interpret them. Any aspect of the construction that does not conform to the requirements set out in Chapter 5 must be regarded as a defect. However, though a defect will make a wall less strong than it might have been, it does not necessarily mean that the wall is no longer fit for its purpose.

A most fundamental defect for a drystone retaining wall is a reduction in its free-draining nature. Pointing a wall will certainly have this effect – mortar pushed into the joints may severely impede drainage of water, even if it seems that drainage pipes have been installed, and gaps left open. The entire volume of a good wall allows water to drain from the retained soil, so that water pressures do not build up; local drainage cannot be as effective. Any wall that has been pointed must be regarded as severely damaged and requires very careful assessment of evidence of effective drainage. If the core of the retaining wall has been made of pieces of stone that are too small, or has had finer material put into it deliberately, then the core may clog, and this can often be checked by inspecting through the gaps between the facing stones.

Figure 6.6 shows two views of a bulged and collapsing section of wall holding back part of an embankment adjacent to a bridge abutment. It can be seen that the wall supporting the bridge, to the left of the left-hand image, is carefully built and in fairly good condition. The bulge is severe, but it is notable that a substantial part of the wall is still standing, even with this extreme bulge. However, this wall is most probably on the verge of a complete collapse. Given appropriate material properties, this could be analysed to demonstrate that fact, but even if such an analysis showed there to be a good margin of stability remaining, the obviously distressed state of the wall would lead to its being dismantled and rebuilt. The more difficult assessment is of a less severe bulge that has been standing for a long time. This is common, and although a full analysis would be needed to give a good level of confidence, a rule of thumb is that if the stones are bulging so that more than a third of their width lies in front of the true plane of the wall face then there are grounds for serious concern, and that if more than half the width does so, then the wall may be on the verge of collapse, as can be seen in Figure 6.6.

Chapter 5 makes clear that a wall should be constructed so that no stone can be extracted from the face, but it is possible for an otherwise well-built

Figure 6.6 A bulging drystone retaining wall in slate. (Courtesy of Professor William Powrie.)

Figure 6.7 A wall adjacent to a highway, in very poor condition with stones falling from the face.

wall to lose stones from the face without problem, and some walls may be observed to have larger areas of the face fallen away, yet the structure still stands. Figure 6.7 shows a wall adjacent to a highway that has been built in *opus quadratum*, yet it can be seen that some of the stones were only squared off on the face, as they are falling away from the wall. Further along this section of road, the wall has almost completely collapsed, leaving a bank of earth and rubble. Figure 6.8 shows part of a test wall at Bath, which has bulged severely and some of the stone has fallen away. Yet the remaining stone spans over the void in corbel fashion, and the wall was still stable at this stage when the concentrated loading at the surface was removed.

Figure 6.8 Part of a test wall at the University of Bath. A section of facing stone has fallen away completely.

Figures 6.9 and 6.10 show sections of wall left standing after a length of wall failed. In Figure 6.9, the wall has been pulled forwards, probably as failure took place, with shearing in the bottom third of the wall, and the top two-thirds remaining approximately upright. In contrast, the section in Figure 6.10 is leaning alarmingly, to the extent that concrete blocks have been placed to prevent it falling, though it is in fact still standing. These sections of wall are useful illustrations because they show the extreme deformations that a wall can undergo without failing – even though they were being dragged by the section that did fail, they still remained standing. Analysis of these sections shows that they would be expected to remain standing, even leaning as far as they do.

Further examples of the effect of defects may be drawn from the full-scale testing work at the University of Bath. Figure 6.11 shows two walls, that on the left built without running joints and that on the right with running joints. In principle, bonding along the length of the wall is not needed to maintain the wall's stability, but if there is a concentrated loading or a local weakness, good bonding allows the load to be redistributed along the length of the wall, or a weakness to be spanned. So a wall that shows running joints may stand safely enough until a heavy vehicle comes close to its top. A well-constructed drystone retaining wall has considerable strength

Figure 6.9 The wall on one side of a failure. Front and back faces can be seen, and the fill between the two has expanded as the front face pulled away.

Figure 6.10 The wall on one the other side of the same failure. The wall has a severe
lean but is otherwise intact. Concrete blocks have been placed in front of
the wall to prevent it from falling over.

in reserve and considerable ductility. So the wall without running joints
was able to carry a heavy localised load, even though it was relatively slender in cross-section, and was able to remain standing even at the extreme
inclination shown. Figure 6.12 shows a test wall built of slate/shale, which
failed by the main part of the wall sliding forwards over its lowest courses,
until it toppled over the edge. A wall may begin to slide, and then come
to a halt if projections on the stone engage with each other sufficiently;
however, if the wall is so close to the limit that it has begun to slide, there
will probably be very little capacity in reserve, and a wall that has clearly
undergone sliding should be regarded as unsafe.

6.3 SUMMARY

Any investigation must begin with a clear understanding of the purpose
of the structure being investigated, which extends to an appreciation of
its significance in its environment. The aims of the investigation must be
clear, and it will be informed by a knowledge of good construction practice for the materials employed in the wall's construction. Methods have

Figure 6.11 Two test walls at the University of Bath. The wall on the left was constructed
to a high standard with no running joints, and as seen in the view below, was
able to maintain its integrity even when extremely deformed. The tensile
strength along the length of the wall, which arose from the well-bonded
construction, allowed the heavily loaded central section to be supported by
the lightly loaded sections on either side. In contrast, the wall on the right
was deliberately built with running joints, along which the wall separated
under concentrated loading.

been described that can give insight into the construction and condition
of a wall without the need for destructive or invasive investigation, but
specialist involvement is essential if investigation proceeds to the stage of
partial dismantling. A wall can remain standing even though it is extremely
deformed, providing that it is carrying a simple and constant earth pressure, but may be very vulnerable to any changes; considerations of future
security are likely to be central to any assessment. Careful construction
following the rules set out in Chapter 5, of a wall dimensioned based on

Figure 6.12 A test wall at the University of Bath constructed from shale. This wall failed in sliding. The targets enable the distance the wall had slid on its base to be seen clearly; this photograph was taken just before the wall failed.

the understanding presented in Chapters 2 through 5, will give a strong and ductile wall. Walls of inferior construction may nevertheless give many years of good service, but may be very vulnerable to changes in their loading, whether brought about by vehicle loading or by heavy rainfall.

Appendix: Design charts – to enable initial sizing

A.1 INTRODUCTION

The design charts presented in this Appendix (Figures A.C1 through A.C18) summarise results of calculations using the limit equilibrium analysis as presented in Section 4.4.1 in Chapter 4. This section relies on the limit equilibrium analysis described in general in Chapter 2, and on the specific behaviour of drystone retaining walls described in Chapter 3. Design charts can be used to provide an indication of the expected geometry for the purposes of initial design of drystone retaining structures. The following design charts were established in 2008 by ENTPE (Ecole Nationale des Travaux Publics de l'Etat) in cooperation with specialised masons in drystone walling and with SETRA (which is now CEREMA). The expertise of such masons is required to ensure compliance with good practice, to produce final constructions that allow the engineering assumption of monolithic behaviour of the walls in 2D. This was verified by experiments (Villemus et al. 2007; Mundell et al. 2009; Colas et al. 2010b, 2012). A comparison between the charts presented here and the Yield design method can be found in the paper of Le et al. (2013).

In accordance with Chapter 3, the limiting condition that is considered involves the upper part of the wall moving, while the lower part remains fixed to the foundation. Therefore, the equilibrium of forces takes into account only the upper part of the wall (Figure A.1). The interface friction angle of the back of the wall and the soil is assumed to be equal to the angle of friction of the soil (Chapter 3).

The safety factor with respect to overturning, which is used for preparing the design charts, is thus calculated on the upper part of the wall (Figure A.1) and equal to 1.5 following the French rules; this is appropriate in the case of drystone retaining walls, given confidence in the parameters, the

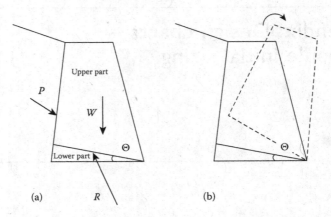

Figure A.1 (a) Forces acting on the upper part of the DSRW. (b) Failure by overturning.

carefully defined mechanism and the ductility of the structures. The case of the failure by sliding (Figure 2.5a) was also calculated for the range of parameters presented in the charts, but overturning was always critical.

A.2 PRESENTATION OF CHARTS

The charts allow a numerical value to be determined without the need for explicit calculation. Each graph represents the results of the same calculation procedure with different parameters. Each curve represents the evolution of the minimum width of the base of a drystone retaining wall in terms of the geometric and physical characteristics of the wall and backfill, considering the stability rules mentioned in Chapters 2 and 3 and Section 4.4.1 of Chapter 4.

A.2.1 Using the charts

The width of the base B given by the charts depends on the features selected or imposed for the wall and backfill (Figure A.2).

A.2.1.1 Parameters of the wall

Height h: Height of the wall, given in metres. The charts provide for walls from 2 to 6 m, in steps of 50 cm.

Front face batter f_1: Angle formed between the outer facing of the wall and the vertical, as a percentage. The charts provide for front batter of 0%, 10% or 20%.

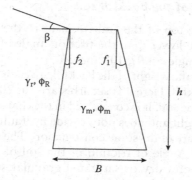

Figure A.2 Geometrical and physical characteristics of the wall and backfill.

Inner face batter f_2: Angle formed between the inner facing wall and the vertical, as a percentage; the charts are all for inner face batter set to zero. In practice, it may be slightly inclined provided that it is always towards the front – the wall must not lean back onto the fill (see Section 5.2.3 in Chapter 5); then the width of the base must be increased beyond that given in the charts to $(B + h \times f_2)$.

Unit weight γ_m: Unit weight of the wall, given in kN/m³; it depends on the self-weight of the stone used in the wall, and on the void percentage in the masonry. The unit weight of the stones is intrinsic to the type of stone used. The percentage of voids relies on empirical and experimental data, and an average value of 25% might be expected if the walls are well built according to the rules of the art given in Chapter 5.

Angle of friction ϕ_m: Friction angle of one stone upon another, given in degrees; it characterises the shear strength of joints of dry stone masonry. This angle can be measured by direct shear tests in the laboratory. In practice, it is possible to put two stone slabs one on the other and incline the lower stone until the upper stone begins to slip; the inclination of the surface on which the upper stone rests is then ϕ_m. Details about the measurement of the stone angle of friction can be found in Ciblac and Morel (2014). For this study, the values obtained by Boris Villemus (Villemus 2004) in his PhD thesis are used. However, the value was only used to check the safety in sliding, which was not used to draw the charts, as explained previously. Caution is advised if the stone proposed is less rough than the limestone and schist considered here, as the sliding may become critical.

The unit weight of the wall and the stones and the friction angle are determined by the type of stone used for construction. The charts are given for two types of stone commonly used and tested by Boris Villemus in his experiments: limestone ($\gamma_m = 16$ kN/m³, $\phi_m = 37°$) and schist ($\gamma_m = 18$ kN/m³, $\phi_m = 28°$).

A.2.1.2 Parameters of the backfill soil

Slope β: Angle of the top of the embankment to the horizontal, given in degrees. This must be lower than the friction angle of backfill. The charts were calculated for angles of 0°, 10° and 20°.

Unit weight γ_r: Specific weight of the backfill, given in kN/m³; it depends on the nature of the backfill. Here it is set arbitrarily at 20 kN/m³ to reflect the presence of water in the soil. It is considered in the charts that this water has a passive role by its weight and does not provide hydraulic pressure because of the free-draining nature of drystone construction. The suction is neglected.

Friction angle ϕ_R: Angle of friction of the soil of the backfill given in degrees. In the case of a dry or saturated granular soil, this angle corresponds to the natural slope angle to the horizontal, found if the material is simply tipped into a pile. However, the presence of moisture and of finer particles makes this behaviour more complex, and the parameter is typically measured in the laboratory by triaxial or direct shear test. This angle is generally between 20° and 40°, but the charts are extended to the range 0°–50° to show the effect of extreme cases. As an indication:

- $\phi_R < 20°$; such a material is not suitable for backfilling a retaining structure
- $\phi_R = 25°$ for clayey soil
- $\phi_R = 30°$ for sandy soil
- $\phi_R = 35°$ for gravel
- $\phi_R = 50°$ for very rare angular and highly frictional soils

It may be noted that the soil cohesion is not considered. This assumption reflects a desire for simplicity of calculations, but is also for safety reasons.

A.2.2 Practical use of charts

The graphs presented in this Appendix represent the change in the minimum width of the base of the wall according to the different parameters mentioned previously. Each chart is chosen for a given stone, wall batter and slope of the backfill. On each chart, there are several areas each corresponding to a different height of wall (2–6 m). Each curve represents the variation of the minimum width of the wall according to the friction angle of the backfill soil.

Before using the charts to design a drystone wall, check the necessary data:

1. Type of stone (limestone or shale)
2. Wall batter
3. Backfill slope
4. Wall height
5. Friction angle of the backfill soil

First, select the chart corresponding to the case within 1–3 (type of stone and backfill slope) and then the curve is determined by the wall height (point 4) and finally yields the desired width of the wall according to the friction angle of the soil (point 5).

Example: to build a schist wall 2.50 m high with a batter of 10% and a sandy backfill with a slope of 10°, refer to the chart that is located in the schist section, batter of wall of 10% and slope angle 10° (Figure A.C14). As the wall height is 2.50 m, the focus is on the third curve from the bottom, so that for $\phi_R = 30°$ (sandy soil), the width of the base of the wall is 1.38 m (Figure A.3).

The guide provides 18 charts in total, corresponding to 2 kinds of stone, 3 slopes of backfill slope (β) and 3 different values of external batter of wall face (f_1) as follows:

1. Materials: limestone, schist
2. Backfill slope: 0°, 10°, 20°
3. External batter: 0%, 10%, 20%

A.3 WARNINGS AND DISCLAIMER

Neither the authors nor the publishers of this book can accept any liability in connection with the use of these charts, which are presented to enable an initial assessment to be made of the likely size of a drystone retaining wall. It is the responsibility of the user to ensure that calculations and checks are carried out, as required in the territory where the wall is to be built, and

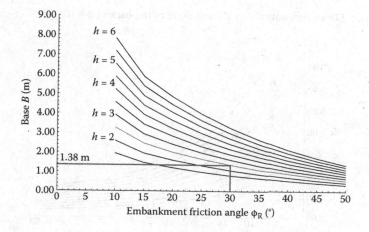

Figure A.3 Determination of the width to the base of a slate wall 2.50 m high with a batter of 10% and a sandy backfill. The chart is the one in Figure A.C14, ϕ_R friction angle = 30° inclined 10°.

bearing in mind the significance of the structure proposed and any legislation and guidelines that must be followed. Such checks and calculations must be carried out by a suitably qualified person, according to the requirements of the location. The practice of civil engineering, and geotechnical engineering especially, requires that a careful assessment is made of the site of a project, as not all hazards are visible, and the appropriate response even to those hazards that are obvious is not always clear. Proper professional involvement at an early stage may be essential to prevent failure, damage to property or even loss of life, whether during construction or at a later date.

A.4 THE CHARTS

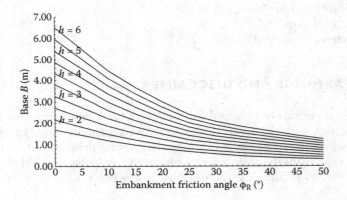

Figure A.C1 Limestone, batter f_i = 0% and slope of the backfill β = 0°.

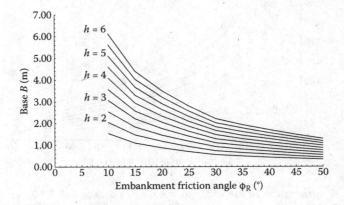

Figure A.C2 Limestone, batter f_i = 0% and slope of the backfill β = 10°.

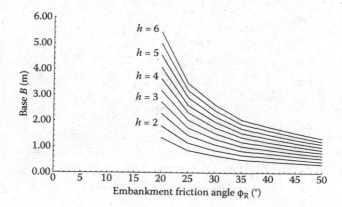

Figure A.C3 Limestone, batter f_i = 0% and slope of the backfill β = 20°.

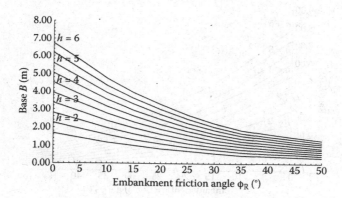

Figure A.C4 Limestone, batter f_i = 10% and slope of the backfill β = 0°.

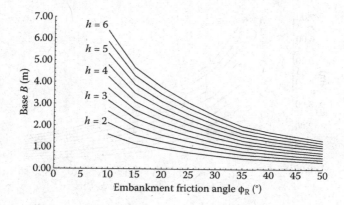

Figure A.C5 Limestone, batter f_i = 10% and slope of the backfill β = 10°.

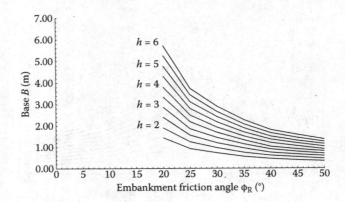

Figure A.C6 Limestone, batter f_1 = 10% and slope of the backfill β = 20°.

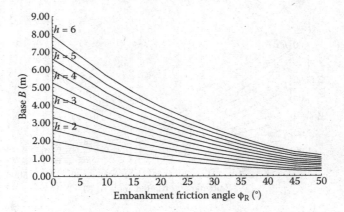

Figure A.C7 Limestone, batter f_1 = 20% and slope of the backfill β = 0°.

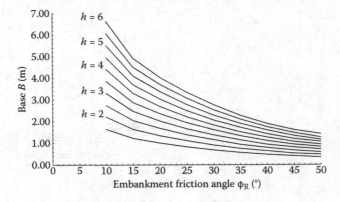

Figure A.C8 Limestone, batter f_1 = 20% and slope of the backfill β = 10°.

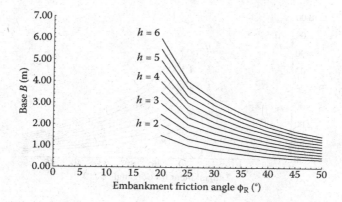

Figure A.C9 Limestone, batter f_i = 20% and slope of the backfill β = 20°.

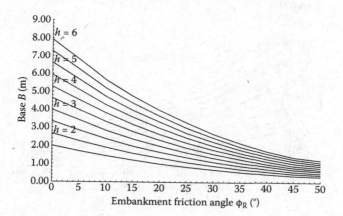

Figure A.C10 Schist, batter f_i = 0% and slope of the backfill β = 0°.

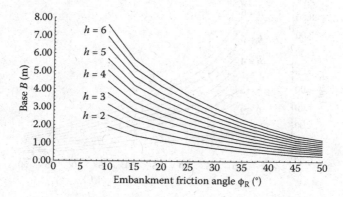

Figure A.C11 Schist, batter f_i = 0% and slope of the backfill β = 10°.

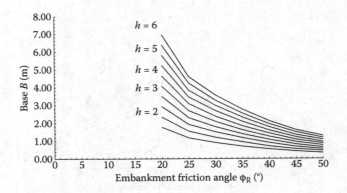

Figure A.C12 Schist, batter f_1 = 0% and slope of the backfill β = 20°.

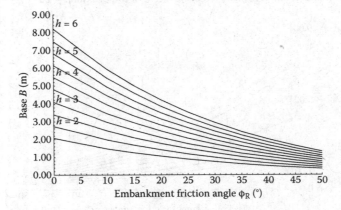

Figure A.C13 Schist, batter f_1 = 10% and slope of the backfill β = 0°.

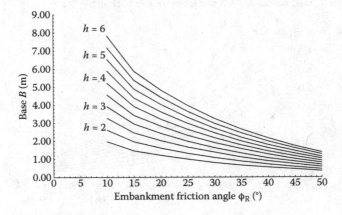

Figure A.C14 Schist, batter f_1 = 10% and slope of the backfill β = 10°.

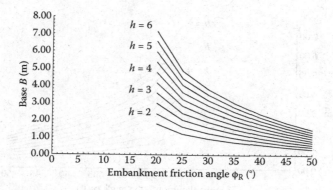

Figure A.C15 Schist, batter f_1 = 10% and slope of the backfill β = 20°.

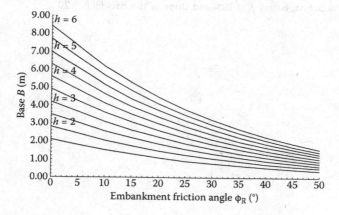

Figure A.C16 Schist, batter f_1 = 20% and slope of the backfill β = 0°.

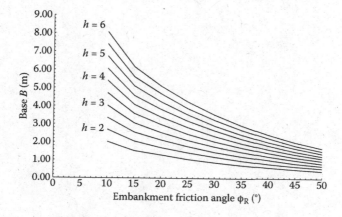

Figure A.C17 Schist, batter f_1 = 20% and slope of the backfill β = 10°.

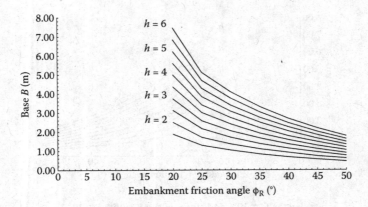

Figure A.C18 Schist, batter f_1 = 20% and slope of the backfill β = 20°.

References

Alava, C., Augeraud, L., Apavou, S., Bouskela, D., Lenoir, C., and Peyrard, M. (2009) Murs de soutènement: Comparaison environnementale et financière de différentes technologies. Projet d'Option, École Centrale de Lyon.

Bastier, R. T. (2000) Cement kiln: Clinker workshops (in French). Techniques de l'Ingénieur, BE8844.

Burgoyne, J. (1853) Revetments or retaining walls. *Corps of Royal Engineering Papers* 3: 154–159.

Confédération de l'Artisanat et des Petites Entreprises du Bâtiment (CAPEB), Artisans Bâtisseurs en Pierres Sèches (ABPS), Murailleurs de Provence, Confrérie des Bâtisseurs en Pierre Sèche (CBPS), Chambre de Métier et de l'Artisanat de Vaucluse (CMA 84), and Ecole Nationale des Travaux Publics de l'Etat (ENTPE) (2008) Guide des bonnes pratiques de construction de murs de soutènement en pierre sèche. Vaulx-en-Velin: ENTPE.

Chen, T. Y., Burnett, J., and Chau, C. K. (2001) Analysis of embodied energy use in the residential building of Hong Kong. *Energy* 26: 323–340.

Ciblac, T., and Morel, J. C. (2014) *Sustainable Masonry: Stability and Behavior of Structures*. Hoboken, NJ: John Wiley & Sons.

Classen, M., Althaus, H. J., Blaser, S., and Scharnhorst, W. (2007) Life cycle inventories of metals. In *Ecoinvent Report v2.0 Report No. 10*. Swiss Centre for Life Cycle Inventories, pp. 81–158.

Claxton, M., Hart, R. A., McCombie, P. F., and Walker, P. (2005) Rigid block distinct element modelling of dry-stone retaining walls in plane-strain. *ASCE Journal of Geotechnical and Geoenvironmental Engineering* 131(3): 381–389.

Colas, A. S., Morel, J. C., and Garnier, D. (2008) Yield design of dry-stone masonry retaining structures – Comparisons with analytical, numerical and experimental data. *International Journal for Numerical and Analytical Methods in Geomechanics* 32: 1817–1832.

Colas, A. S., Morel, J. C., and Garnier, D. (2010a) 2D modelling of a dry joint masonry wall retaining a pulverulent backfill. *International Journal for Numerical and Analytical Methods in Geomechanics* 34: 1237–1249.

Colas, A. S., Morel, J. C., and Garnier, D. (2010b) Full-scale experiments to assess dry stone earth retaining wall stability. *Engineering Structures* 32: 1215–1222.

Colas, A. S., Brière, R., Feraille, A., Habert, G., and Tardivel, Y. (2014a) Holistic approach of a new masonry arch bridge on a Cevennes road. In *9th International Masonry Conference 2014*, July 7–9, 2014, Guimarães, Portugal.

Colas, A.-S., Garnier, D., Habert, G., Tardivel, Y., and Morel, J.-C. (2014b) Advances on structural, environmental and economical analysis of drystone retaining walls. In F. Peña and M. Chávez (Eds.), *SAHC2014 – 9th International Conference on Structural Analysis of Historical Constructions*, October 14–17, 2014, Mexico City, Mexico.

Constable, C. (1874) Retaining walls – An attempt to reconcile theory with practice. *American Society of Civil Engineers* 3: 67–75.

Cooper, M. (1986) Deflections and failure modes in drystone retaining walls. *Ground Engineering* 19(8): 28–33.

Cundall, P. A. (1971) A computer model for simulating progressive large scale movements in blocky rock systems. In *Proceedings of the Symposium of the International Society for Rock Mechanics*, Paper No. II-8.

de Buhan, P., and de Felice, G. (1997) A homogenization approach to the ultimate strength of brick masonry. *Journal of the Mechanics and Physics of Solids* 47(7): 1085–1104.

Fédération Nationale des Travaux Publics (FNTP) (JCB JS 180 LC AMS): http://fr .equipment-center.com/cote/materiels-par-type-et-marque/pelles-hydrauliques-sur -chenilles_d100/jcb.htm (Accessed 1 April 2010).

Frischknecht, R., Jungbluth, N., Althaus, H., Doka, G., Dones, R., Hischier, R., Hellweg, S., Humbert, S., Margni, M., Nemecek, T., and Spielmann, M. (2007) Implementation of life cycle impact assessment methods: Data v2.0. In *Ecoinvent Report No. 3*. Swiss Centre for Life Cycle Inventories, Dübendorf.

Gupta, V. P., and Lohani, N. K. (1982) Treatment and repair of partially damaged retaining walls in hills. *Indian Highways* 10(10): 20–28.

Habert, G., Castillo, E., and Morel, J. C. (2010) Sustainable indicators for resources and energy in building construction. In *Second International Conference on Sustainable Construction Materials and Technologies*, June 28–30, 2010, Ancona, Italy.

Habert, G., Castillo, E., Vincens, E., and Morel, J. C. (2012) A new energy indicator for life cycle analysis of buildings. *Ecological Indicators* 23: 109–115.

Harkness, R. M., Powie, W., Zhang, X., Brady, K. C., and O'Reilly, M. P. (2000) Numerical modeling of full-scale tests on drystone masonry retaining walls. *Géotechnique* 50(2): 165–179.

Heyman, J. (1966) The stone skeleton. *International Journal of Solids and Structures* 2(2): 249–279.

Heyman, J. (1988) Poleni's problem. *Proceedings of the Institution of Civil Engineers*, Part 1, 84(August): 737–759.

Joint Research Centre (JRC) (2000) Integrated pollution prevention and control: Reference document on best available techniques in the cement and lime manufacturing industries. European Commission, Bruxelles.

Kawai, K., Sugiyama, T., Kobayashi, K., and Sano, S. (2005) Inventory data and case studies for environmental performance evaluation of concrete structure construction. *Advances in Concrete Technology* 3: 435–456.

Le, H. H., Morel, J. C., Garnier, D., and McCombie, P. (2013) A review of design methods of dry-stone retaining walls, *Geotechnical Engineering* 167(3): 262–269.

Marcom, A. (2002) Enquête sur la mesure de productivité dans les techniques de construction en terre. DPEA Terre, Ecole d'Architecture de Grenoble.

Marcom, A., Floissac, L., Colas, A. S., Bui, Q. B., and Morel, J. C. (2009) How to assess the sustainability of building construction processes. In *Fifth Urban Research Symposium – Cities and Climate Change: Responding to an Urgent Agenda*, June 28–30, 2009, Marseille, France.

Martaud, T. (2008) *Evaluation environnementale de la production de granulats naturels en exploitation de carrière: indicateurs, modèles et outils*. Ph.D. thesis. Orléans University, France, 212 pp.

McCombie, P. F., Mundell, C., Heath, A., and Walker, P. (2012) Drystone retaining walls: Ductile engineering structures with tensile strength. *Engineering Structures* 45: 238–243.

Morel, J. C., Mesbah, A., Oggero, M., and Walker, P. (2001) Building houses with local materials: Means to drastically reduce the environmental impact of construction. *Building and Environment* 36: 1119–1126.

Mundell, C., McCombie, P., Bailey, C., Heath, A., and Walker, P. (2009) Limit – Equilibrium assessment of drystone retaining structures. *Proceedings of the Institution of Civil Engineers Geotechnical Engineering* 162(GE4): 203–212.

Norme Française European Norm (NF EN) International Standard Organisation (ISO) 14040 (2006) *Environmental Management – Life Cycle Assessment – Principles and Framework*. Saint-Denis La Plaine: Association Française de Normalisation.

NF EN ISO 14044 (2006) *Environmental Management – Life Cycle Assessment – Requirements and Guidelines*. Saint-Denis La Plaine: Association Française de Normalisation.

Odent, N. (2000) Recensement des ouvrages de soutènement en bordure du réseau routier national. *Ouvrage d'Art* 34: 15–18.

Omer, A. M. (2008) Energy environment and sustainable development. *Renewable and Sustainable Energy Review* 12: 2265–2300.

O'Reilly, M. P., Bush, D. I., Brady, K. C., and Powrie, W. (1999) The stability of drystone retaining walls on highways. *Proceedings of the Institution of Civil Engineers, Municipal Engineering* 133(2): 101–107.

Powrie, W. (1996) Limit equilibrium analysis of embedded retaining walls. *Géotechnique* 46(4): 709–723.

Rigassi, V., and Seruzier, M. (2002) Bilan économique, social et environnemental de 20 ans de filière blocs de terre comprimée à Mayotte, Direction de l'Équipement/ Société immobilière de Mayotte, Juillet.

Salençon, J. (1983) *Calcul à la rupture et analyse limite*. Presses de L'Ecole Nationale des Ponts et Chaussées, Paris.

Salençon, J. (2013) *Yield Design*. Mechanical Engineering and Solid Mechanics Series, ISTE. Hoboken, NJ: John Wiley & Sons, London.

Villemus, B. (2004) Etude des murs de soutènement en maçonnerie de pierres sèches, Thèse de doctorat de l'Institut National des Sciences Appliquées et de l'Ecole Nationale des Travaux Publics de l'Etat, 247 pp., Lyon.

Villemus, B., Morel, J. C., and Boutin, C. (2007) Experimental assessment of dry stone retaining wall stability on a rigid foundation. *Engineering Structures* 29(9): 2124–2132.

Walker, P. J., and Dickens, J. G. (1995) Stability of medieval dry stone walls in Zimbabwe. *Géotechnique* 45(1): 141–147.

Walker, P., McCombie, P., and Claxton, M. (2007) Plane strain numerical model for drystone retaining walls. *Proceedings of the Institution of Civil Engineers, Geotechnical Engineering* 160(GE2): 97–103.

Index

Page numbers ending in "f" refer to figures. Page numbers ending in "t" refer to tables.

Printed in the United States
by Baker & Taylor Publisher Services

Printed in the United States
by Baker & Taylor Publisher Services